The Deceptive Brain

Blame, Punishment, and the
Illusion of Choice

The Deceptive Brain

Blame, Punishment, and the Illusion of Choice

Robert L. Taylor, M.D.

IFF
BOOKS

Winchester, UK
Washington, USA

JOHN HUNT PUBLISHING

First published by iff Books, 2021
iff Books is an imprint of John Hunt Publishing Ltd., No. 3 East Street, Alresford,
Hampshire SO24 9EE, UK
office@jhpbooks.com
www.johnhuntpublishing.com
www.iff-books.com

For distributor details and how to order please visit the 'Ordering' section on our website.

ISBN: 978 1 78904 755 4
978 1 78904 756 1 (ebook)
Library of Congress Control Number: 2020945942

A CIP catalogue record for this book is available from the British Library.

Design: Stuart Davies

UK: Printed and bound by CPI Group (UK) Ltd, Croydon, CR0 4YY
Printed in North America by CPI GPS partners

We operate a distinctive and ethical publishing philosophy in
all areas of our business, from our global network of authors to
production and worldwide distribution.

Contents

Other Books by this Author

Mind or Body: Distinguishing Psychological from
Organic Disorders
ISBN: 978-0-0706-2963-9

Health Fact, Health Fiction: Getting Through the Media Maze
ISBN: 978-0-8783-3683-8

Psychological Masquerade: Distinguishing Psychological from
Organic Disorders, Third Edition
ISBN: 978-0-8261-0247-8

Finding the Right Psychiatrist: A Guide for
Discerning Consumers
ISBN: 978-0-8135-6624-5

Madhouse Blues
ISBN: 978-1-6355-4025-3

To
Gemma 'Nuni'

and

Drexel

Emma

Finley

Grace

Jack

James

Mallory

Masen

Weston

May the stories you live be engaging, satisfying, and most of all full of wonder.

Preface

We are not who we seem to be. Not even close. Central to our understanding of human experience is an unswerving conviction in a mind that puts us in control and allows us to choose and will our way through life. Despite this deep-seated universal belief, rigorous attempts to prove it have met with surprising failure.

In the 1930s-50s the innovative American-Canadian neurosurgeon Wilder Penfield greatly expanded the scope of brain surgery with a remarkable series of studies on patients with various forms of epilepsy (Gilder, 1989). In addition to exploring effective treatments, Penfield was intent on finding the mind's location in the brain. Over a period of 30 years, with the imaginative use of an electronic probe, he conducted his search on over 1,000 patients. At surgery his subjects were fully alert and without any pain. By selectively stimulating various parts of the brain, Penfield was able to elicit involuntary speech, memories, and various motor movements. Fully conscious subjects marveled at having these things happen outside their control. But in the end Penfield's search for the mind proved fruitless. Its most critical elements—deciding, willing, and imagining—were nowhere to be found (Penfield, 1975).

More recently this search for the mind was taken up by Francis Crick. Crick was a man of great curiosity, an intellectual heavyweight. He was also a wanderer. After a distinguished career in physics, he turned to biology and what he considered one of the two most important scientific questions: *What is the physical basis of life?* In his mid-thirties he collaborated with a much younger James Watson, and in a feverish race to decipher the structure of DNA—the basic unit of genes—the two men (along with a mighty assist from Rosalind Franklin) bested world famous organic chemist Linus Pauling. Their discovery garnered both men Nobel Prizes and gave birth to biotechnology.

But it was not enough for Crick. He grew restless again and after 30 years of molecular biology left Cambridge for a new academic home at the University of California, San Diego. Using the brilliant minds around him to teach himself neuroscience, Crick set to work trying to answer the second of his two great questions, *What and where is consciousness?* (No one ever accused Crick of being modest in his goals.)

How is it possible, he asked, that within a second of viewing an object we experience full color, 3-dimensional vivid sight? How do electrochemical events in the brain burst forth as something entirely different? How does light on a retina eventually become a rose? In an attempt to answer this question, Crick and his colleagues engaged in a detailed exploration of the brain's visual system, mapping how multiple brain inputs "bind" together. But despite years spent meticulously tracking light as it made its way from the retina through nerves, neural tracks and nuclei to the visual cortex at the back of the brain and eventually to higher level associative centers, Crick failed to find what he was really looking for. The "soul" (his word for mind) remained stubbornly beyond his reach. There would be no second Nobel. Finally, in his book, *The Astonishing Hypothesis*, he was forced to conclude: "You, your joys and your sorrows, your memories and your ambitions, your sense of personal identity and free will are in fact no more than the behavior of a vast assembly of nerve cells... As Lewis Carroll's Alice might have phrased it: 'You're nothing but a pack of neurons'" (Crick, 1994).

Crick died at age 88 still holding to his cynical conclusion: human experience is all neurons, nothing more. But obviously there is more. Crick and Penfield's failures to find the mind in the brain hardly mean it doesn't exist. Although mystery remains as to how biological stuff in the brain emerges as mindful experience, it is something we vividly experience every day—unexplained but as real as anything we know. (In fact, it's all we know.) The colors we see, the taste and smell of food, the experiences of love, anger, defeat, and triumph as well as all the things we imagine are not to be dismissed simply because scientists can't find them in the brain.

Still, the original question about the mind has remained unresolved. How does the electrochemical action of millions of microscopic neurons balled up in a 3-pound brain give rise to who we are and what we experience? How is it we seem free to choose our way through our lives while everything around us occurs in strictly determined, cause-and-effect fashion?

The Deceptive Brain picks up where Crick and Penfield left off. They failed in their quest by limiting their search to a material mind, overlooking the possibility that our minds might be something entirely different. Neither of them considered the possibility of mind as an emergent phenomenon that has its own subjective reality in which human experience takes place. (More about this later. Much more.) Over eons of time, in a deceptive but highly creative maneuver, the brain has engineered a decidedly unbrain-like product: a narrative translation. Instead of the actions of neurons and electrochemical reactions, at the heart of this after-the-fact story is the self—I and me—as chief protagonist.

While I will readily acknowledge that on first glance this conclusion seems spooky and preposterous, I can assure you it is not. In fact it is far better substantiated than the conviction each of us wakes up with every morning sensing we are free-willing entities defying basic laws of cause and effect as we willfully choose our way through life. Although details remain to be filled in, the outline of this highly counterintuitive view of human experience (based on compelling neurobehavioral science evidence) is already firmly in place. If true, this alternative version of who we are raises profound questions, particularly with respect to blame and punishment. One of the most counterintuitive implications being that *no one is ever to blame for anything*. Responsible, yes, but not blameworthy. If human experience is an after-the-fact orienting story where we act out fully determined decisions already made, blame and punish are misdirected, if not immoral.

Much of the early part of this book deals with the foibles of trying to assign blame and punishment fairly and consistently. In the absence of an awareness of the illusion of choice, errors commonly made in the conduct of "justice" are understandable. But that was then; now is now. The evidence of a different reality is there for anyone to see who wants to see it.

I have tried to rescue this subject from the tightly-held grip of philosophers and theologians where it has resided for centuries by placing it in the real-world context of criminal justice. For too long this counterintuitive view of choice and free will has been absent from discussions regarding its reform. It now deserves a seat at the table.

Introduction: Waning of *Homo Grandiose*

God's noblest work? Man.
Who found it out? Man.
Mark Twain

As humans, we have proclaimed ourselves king of the hill for quite some time, but a relentless accumulation of contradictory evidence has gradually undermined this claim. The portrayal of man as the center of things—the main attraction of the universe, second only to God, of unusual intelligence and design—has been badly bludgeoned by a slow-moving series of revelations.

Move from High to Low Rent District

It started with the discovery that our home—earth—is *not* the center of things but only a lowly planet circling a second-rate sun located in a massive universe. After years of astronomical observations and mathematical calculations, the Renaissance mathematician and astronomer Copernicus made this startling finding which initially he was reluctant to divulge for fear of the outrage it might incite. Rumors had already spread across Europe before he finally published (on the same day he died in 1543) his mind-shattering discovery: the sun—not the earth—was the center of the world. Oddly enough, resistance to the idea was slow to develop. More than six decades passed before the Catholic Church took on the Copernican challenge, arguing feebly that astronomical findings were nothing more than intellectual abstractions with no real import. (An early version of *fake news*.) But by that time it was too late. The truth was out. In the overall scheme of things, earth—man's home—was not what it seemed, the center of things.

5

Big Come Down

In the middle of the 19th century, a man slated earlier to be a minister got the opportunity to travel the world as a naturalist. It was on this trip that Charles Darwin began to understand how all life evolves through a process of adaptation and selection. As Darwin would eventually conclude, humans were not a special creation but merely one of many of life's numerous product lines. Like Copernicus, he agonized over the religious blow back he knew his radical thesis would generate. Even after 30 years of pulling together and refining his thoughts on evolution, Darwin still delayed publication. It took delivery of a short paper by Alfred Russel Wallace from halfway around the world outlining similar ideas to finally galvanize him into action. Eventually, both men were recognized, but Darwin's elegant and exhaustive account in *On the Origin of Species* made his name forever synonymous with evolution and the extraordinary conclusion that if we went back long enough in time we would find a common ancestor for *all* life. In his book, *Homo Deus: A Brief History of Tomorrow*, Yuval Noah Harari illustrates the point a different way: "Just 6 million years ago," he says, "a single female had two daughters. One became the ancestor of all chimpanzees, the other is our own grandmother." Harari is breaking the news to us easy when he fails to mention how our line goes all the way back to single cell organisms.

The fact that we share a variety of genes with other life forms confronts us with a host of surprising relatives. Evolutionary scientists tell us *all plants, animals (including humans), and fungi (mushrooms) share a common ancestor;* one who lived roughly 1.6 billion years ago. Eight to ten percent of human DNA originated in viruses! Twenty-five percent of our genes we share with rice, 61% with fruit flies. Ninety-seven percent plus of our genome is the same as that found in orangutans; close to 99% in chimpanzees (*National Geographic*, 2013). In the 1980s geneticists studying flies discovered a group of genes they called *hox* genes, which served as an instruction manual for how to assemble the various parts: head, legs, wings, and so on. The surprise in the scientific community was matched only by a subsequent discovery of these identical hox genes doing the same thing in mice. Subsequently, a string of similar studies forced a stunning conclusion: "... the basic body plan of all animals had been worked out in the genome of a long-extinct ancestor that lived more than 600 million years before and had been preserved ever since in its descendants (and that includes me and you)" (Ridley, 2004). It's all quite disturbing for proponents of human exceptionalism.

Code Deciphered

All we now know about genes we owe to an unassuming, chronically anxious man who on trips to visit the sick or dying was so stressed he sometimes took to bed. Gregor Mendel was an Austrian monk who sought out quiet places perfect for gardening (Henig, 2000). But it wasn't an interest in food production that drove him. He wanted to know why common peas change in character from one crop to the next. With this objective in mind, he planted thousands of pea plants (including numerous varietals) in the monastery garden of St. Thomas' Abbey and then meticulously took notes on seven different traits: seed shape, flower color, seed coat tint, pod shape, unripe pod color, flower location and plant height. Based on six years of observation Mendel puzzled his way through how these different traits were inherited. He saw how they passed from parent plants to their offspring in predictable fashion and how certain ones would even skip a generation before reappearing. He worked out how some were dominant in their inheritance pattern while others were recessive.

It took a while for anyone to notice. After Mendel published his findings with little fanfare, it took more than 30 years before they were rediscovered in 1900 (Mendel had been dead for 16 years). The discipline of genetics was born. His meticulous studies opened the door to an understanding of how human behavior commonly perceived as emanating mainly from self-willed action was in fact hugely affected by genes. (It would take half a century for James Watson, Francis Crick and Rosalind Franklin to work out the precise chemical structure of genes and open the door to modern genetic research.) Today, it's not uncommon for studies of various human behaviors and traits to show 40% or higher directly related to genetics. One study from the Minnesota Center for Twin & Family Research explored the causes of happiness. What they found was unexpected. While a number of factors such as educational level, family income, marital status, and religious commitment contributed less than 3% each, genetics accounted for a whopping 44-52% (Lykken, 2018).

Mind Under the Microscope

Coming from an entirely different place, at the turn of the century, the Austrian neurologist Sigmund Freud became convinced that impulses and motives *beyond our conscious awareness and control* explained much of human behavior. For Freud the pervasive human claim of being captain of one's fate was pure fantasy. Eventually, he would lose his way in the "weeds" of convoluted psychoanalytic assertions, more circular than helpful, but his basic contention survived: just below the surface humans are more like other animals than different, driven by powerful unconscious factors (Storr, 1989).

Freud was followed by others who identified further constraints on human "freedom." B.F. Skinner, a behavioral psychologist as well as a novelist and social philosopher, took an entirely different tact. What you couldn't see or measure— such as the unconscious—was not important, he insisted. Understanding human experience was simply a matter of observing behavior and watching to see what encouraged it and what suppressed it. At its core, human behavior was the product of rewards and punishments. Life was filled with M&M's and switches. Enough said. In his most famous book, *Beyond Freedom & Dignity*, Skinner put it this way: "As a science of behavior adopts the strategy of physics and biology, the autonomous agent to which behavior has traditionally been attributed is replaced by the environment—the environment in which the species evolved and in which the behavior of the individual is shaped and maintained."

Other social scientists emphasized different "invisible" influences exerted by social and economic environments. Karl Marx, the father of communism, and Émile Durkheim, the famous French sociologist, both recognized the destructive effects of divisions of labor and economic inequities. The Austrian zoologist and ethnologist, Konrad Lorenz, known best for his description of parental imprinting in birds, saw aggression as one of the most powerful innate impulses throughout the animal world and claimed humans were no exception. Lorenz lived to see two horrendous world wars that seemed to validate his theory.

More recently, social psychologists have demonstrated the power of groups (group think) over our decision making. Studies show we perceive lines of obviously different lengths as being the same when others say it's so and routinely distort new information through the lens of what we have previously believed (confirmation bias). Faced with massive amounts of data flooding our lives through social media we grow susceptible to oversimplified explanations designed to influence our opinions (Kahneman, 2011). Recent experience with "fake news" drives home the point. Our image of independent minds meticulously distinguishing truth from fiction for ourselves has undergone a notable tarnishing. In truth we are far more susceptible to outside influences than we think.

A Man and Three Women: Blurring the Boundaries

Louis Leakey's career as a paleontologist was sensational. His work excavating the Olduvai Gorge in Tanzania produced remarkable finds including the discovery of *Homo Habilis*, a primate forerunner of modern man. But equally important was his sponsorship of two unlikely young women who went on to do courageous and exceptional work exploring face to face our close relatives, chimpanzees and gorillas.

Jane Goodall left her British school when she was 18 and took work as a secretary, but secretly she dreamed of going to Africa. In 1957 she visited a friend in the Kenyan Highlands and on a lark called Louis Leakey. Surprisingly, it turned out she had developed quite an interest in chimpanzees and asked if she might meet him. Something in the inexperienced Goodall inspired Leakey's trust. He sent her to London for an intensive course in primate behavior and then raised funds for her to go to the Gombe Stream National Park in Tanzania, accompanied by her mother (at the time a Park requirement). The rest is incredible history.

Through persistence and courage, Goodall earned the trust of a troop of chimps. Eventually she was joined by her new husband, Hugo van Lawick, a premier naturalist photographer. Together they chronicled the lives of chimps as it had never been done before. At a time when humans were still considered unique because of their ability to use tools, Goodall was the first to show this was not true when she recorded chimps breaking twigs off trees, stripping them of leaves, and using them to extract termites for food. She also captured in great detail the workings of a chimp community; the social support, hugs, kisses, grooming, and later the darker side of group violence— all of which showed striking similarities to humans (Goodall, 1988).

Leakey also sponsored Dian Fossey. They met after she had taken out a sizeable bank loan to go on an African safari to view mountain gorillas. When she arrived at Olduvai Gorge, Leakey agreed to sponsor her to study these same gorillas in the wild, at the time seemingly on their way to extinction. (Before she left, Leakey insisted Fossey have her appendix out to eliminate the risk of appendicitis when she was isolated in the African jungle.) Shortly afterwards Fossey took up residence in a remote Volcanoes National Park (Rwanda) mountain cabin from which she trekked in to live among gorillas.

In contrast to the public image of King Kong, the impression of gorillas Fossey chronicled was one of curious and affectionate social beings. In a 1971 video, she recorded her favorite, a young male gorilla exploring himself in a mirror as "he began to twist his head back and forth like a teenager primping for the prom" (Strochlic, 2017). In a highly-charged political atmosphere, Fossey fought to protect her gorillas until her luck finally ran out. Tragically, she was murdered in her cabin, December 26, 1985 (Bouton, 1983). The crime was never solved.

Later, following in the footsteps of these two pioneering women primate researchers, Susan Savage-Rumbaugh, psychologist and primatologist, working up close and personal with two bonobos, made a discovery that led to the design of a remarkable research center outside Des Moines, Iowa. There are no cages in this 18-room compound where bonobos live in rooms connected by corridors and hydraulic doors they can open themselves. The compound includes a music room with drums and keyboard. There are blackboards with chalk and a greenhouse supplied with bananas and sugarcane. Woven through the compound is a pervasive emphasis on the bonobo residents caring for themselves. Included is a custom-designed kitchen, a snack room with vending machines and a television with DVDs. But by far the most surprising feature are the touchscreen keyboards containing more than 300 pictorial symbols (for English words) located in each room which allow the bonobos to communicate with their human caretakers.

Based on her extensive experience and research Savage-Rumbaugh concluded that despite the pervasive belief that only humans can use language she "just happened to find out it wasn't true" (Stern, 2020).

Goodall, Fossey, and Savage-Rumbaugh have given us undeniable evidence that our primate cousins are far more human-like than what had been long assumed. One more blow to human exceptionalism.

No More Brightest Kid on the Block

Automata were the forerunners of modern artificial intelligence and robots. In the golden age of Persia (15th century)—a period of great discovery, too often neglected—the three Banu Musa brothers published *The Book of Ingenious Devices* filled with intriguing mechanical creations (Ford, 2015). What they left out was their own extraordinary programmable device. *The Instrument Which Plays by Itself* was a water-powered organ whose notes were produced by a "pinned" cylinder barrel with small teeth-like projections tripping a series of levers that opened and closed the organ pipes. A thousand years before the first computers, the brothers created a vinyl record with music embedded in a rotating wax-covered drum for playback on a self-playing instrument. They called it "cutting" a melody. The organ could be preset to perform a variety of melodies. For an extra touch of mystery the brothers came up with a way of encasing the instrument in a human-like automaton to give the illusion that it was doing the playing. But as ingenious as the Persian automata were, they retained a distinctly mechanical appearance.

AI and Robotics

This all changed in the second half of the last century when challenges to the concept of "homo grandiose" arose in the form of sophisticated computers, artificial intelligence, and robots.

The speed of these advances has been breathtaking. We now have artificial intelligence supporting doctors, robots performing surgery, cars driving themselves, voices (Siri, Echo, Alexa, etc.) informing us from our cell phones and executing household directions. In 2005 a self-driving vehicle completed a grueling desert off-road course designed to challenge autonomous cars (*TIME*, 2017). Now a future of self-driving cars is assumed.

An IBM computer—*Deep Blue*—beat world chess champion Russian Gary Kasparov, only to be superseded by the question-answering computer, *Watson*, which handily vanquished two long-running *Jeopardy!* champions. *Watson* has now become a consultant to numerous enterprises, most notably medicine where its analytic abilities exceed medical experts. A Google AI program diagnoses more than 50 eye diseases better than many clinical experts (Susskind, 2020).

IBM's latest "brainiac" has been tagged *Project Debater*. Recently in a live San Francisco debate against two skilled human debaters (questions dealt with the advisability of government's subsidizing space exploration and the worthiness of telemedicine) AI held its own. Each side gave a four-minute introduction, a four-minute rebuttal, and a two-minute closing statement. According to a crowd of journalists, *Project Debater*, while slightly outscored in the first debate, came out the winner on the telemedicine question and even managed a passable joke in the process. When the opposing debater emphasized the importance of "the human touch," Debater quipped: "I am a true believer in the power of technology, *as I should be*" (Bohn, 2018).

The challenges to human intelligence keep rolling in. Google's machine, *AlphaGo* (developed by London-based *DeepMind*) defeated Fan Hui, a champion player of the ancient board game of GO, five games to none! In this 2016 match up *AlphaGo's* training consisted of reviewing thousands of previous contests. One year later a newer version (AlphaGo Zero)—given only the rules of *Go without* access to any previous contests—achieved world class proficiency in a single day. This same self-teaching machine attained Kasparov-level chess competence in a matter of hours (Rees, 2018).

Aero Researchers at FACEBOOK, using a neural network architecture modeled on the brain, created a program capable of recognizing faces with 97% accuracy. In January 2017, *Libratus*, a computer program designed at Carnegie Mellon—beat four of the world's best players in poker (Texas Hold 'Em), and in early 2018 two AI reading systems developed by Alibaba and Microsoft triumphed in head-to-head contests against educated humans. On a reading comprehension test developed by Stanford (SQUAD) scores were close, but both AI systems bested their human counterparts (*USA Today*, 2018).

It's hard to ignore the "human-like" quality of many newly-created software productions. Take for example *StatsMonkey*. Without human assistance this program analyzes data from a particular sporting event and then with lightning speed composes an attention-grabbing "reporter-like" story highlighting the most decisive plays and players. This amazing piece of software has given birth to a whole new company, *Narrative Science*, churning out news stories for many top media outlets passed off as regular reporter creations (Ford, 2015). In February 2019, the nonprofit *Open AI* announced a new AI system so good at composing text the developers have refused to release it due to its extraordinary potential abuse. Reportedly, this system can compose intelligent-sounding responses to mere prompts and can compose convincing fantasy prose and fake celebrity news.

Although still early in its development, *robotics* is already breathing down the neck of human performance. Twenty years of Japanese research has produced what is called the *Hybrid Assistive Limb (HAL)*, a battery-powered "hard" exoskeleton suit outfitted with numerous sensors. When a person with mobility problems puts it on, *simply thinking about standing or walking* triggers brain impulses that turn on powerful motors to provide mechanical assistance which allows the person to walk.

A California start-up—*Momentum Machines, Inc.*—is working on the *total* automation of a gourmet hamburger shop. The machine does it all: shapes the patties, grills them to order, toasts the bun, slices the added ingredients (lettuce, tomato, pickles, etc.). All of this only after a specific order has been placed. The burgers come out individually customized and completely assembled. Describing the ultimate objective, the co-founder of the company says: "Our device isn't meant to make employees more efficient. It's meant to completely obviate them."

Currently, *machines that learn* are giving AI researchers insights into how the mind evolves (Kwon, 2018). Not long before he died, Richard Feynman wrapped up a stellar career in theoretical physics with this conclusion: "What I cannot understand, I cannot create." In a book entitled, *Exploring Robotic Minds*, Jun Tani turns the idea around: "I can understand," he argues, "what I can create." He insists the best way to understand the human mind is to make one (Tani, 2017). The way the field is progressing, it seems within the realm of possibility.

David Levy, an AI expert and author of *Love + Sex with Robots: The Evolution of Human-Robot Relationships*, claims if you think of Siri (of smartphone fame) as the future of robots, you are lowballing. He predicts the evolution of the intellectual and emotional capacities of robots will be astounding. "They will look like humans (or however we want them to look). They will be more creative than most creative humans. They will be able to conduct conversations with us on any subject, at any desired level of intellect and knowledge, in any language, and any desired voice—male, female, young, old, dull, sexy. The robots of the mid-twenty-first century will also possess human-like or superhuman-like consciousness and emotions" (Levy, 2007).

Ten years ago, inventor and author, Ray Kurzweil, pointed out how the ratings of the best chess computers were improving by 45 points a year while the skill of the best human players remained relatively unchanged. Based on this trend, he made what at the time appeared an outlandish prediction when he identified 1998 as the year when a computer would triumph over a chess champion. As it turned out, he was off by one year. In his best-selling book, *The Age of Spiritual Machines*, Kurzweil states human and computer intelligence will be virtually equal by the third decade of this millennium at about the same time learning machines begin to take on human characteristics. "The machines will convince us that they are conscious, that they have their own agenda worthy of our respect. We will come to believe that they are conscious much as we believe that of each other. More so than with our animal friends, we will empathize with their feelings and opinions. They will embody human qualities and will claim to be human. And we'll believe them." Kurzweil is convinced, these advanced machines will experience a sense of willful self-determination (Kurzweil, 1999).

As to the darker side of AI, a recently developed political tool known as *deepfake* is a product of advanced machine learning. It makes possible the creation of realer-than-real appearing videos of anyone—including celebrities, politicians, and other powerfully influential persons—in which the person appears to say outrageously fake statements. Even more disturbing, this deepfake technology is in its infancy. What's to come likely will be even more disturbing. Not the kind of thing we want to see in an increasingly polarized society.

Building as it has for centuries, the slow-moving assault on *human exceptionalism* is accelerating. Scientists have constructed a "24-hour clock" of earth's history. Starting with its birth approximately 4.5 billion years ago (12 midnight), elemental life doesn't appear until about 4:00 o'clock in the morning. All other forms of life from simple to the most complex roll out through the rest of the day and evening until humans finally arrive on the scene just before midnight at 11:59! Humbling.

We are definitely new kids on the block, part of a history that strongly suggests we will be superseded.

What's Left?

All of this brings us to the subject of free will. Despite the relentless erosion of human exceptionalism, one characteristic has remained largely unchallenged: *self-agency* by which we choose and will our way through life. Reputedly, this unique human capacity allows us to rise above other influences and freely select what actions we take or don't take.

Although this is a deeply ingrained belief, making the logical case for it has always been a tough intellectual slog. In fact we believe it not so much as a matter of logic and evidence but more due to the vividness of our own personal experiences. We decide we want a coffee, tea, or juice. We get up and get it, all so obviously under our control. Open and shut case. Short of being coerced, we never doubt we are free agents. It's just the way things are. Something we don't question. We know it's true because we experience it as true every waking hour of our lives.

But neurobehavioral science evidence is accumulating that strongly suggests we may be deluding ourselves. The late Tom Wolfe, chronicler extraordinaire of the shifting tides of the human condition, summed it up this way: "... the notion of a self who exercises discipline, postpones gratification, curbs the sexual appetite, stops short of aggression and criminal behavior—a self who can become more intelligent and lift itself to the very peaks of life by its own bootstraps through study, practice, perseverance, and refusal to give up in the face of great odds—this old-fashioned notion (what's a bootstrap, for God's sake?) of success through enterprise and true grit is already slipping away..." (Wolfe, 1997). The role of free will— previously debated mainly by philosophers and theologians— is now getting wider attention. Its validity or lack thereof has far-reaching implications particularly for how we treat rule breakers and criminals, and that is why it has become a concern for the nascent discipline of *neurolaw* (Petoft, 2015; Greene, 2004).

If we put aside our experience for a moment and drill down on the idea itself that we choose and free will ourselves through life, it seems out of sync with everything else around us. For it to be true we have to believe in an entity—a genii-like agent—residing somewhere inside us that routinely acts spontaneously (magically) as it chooses and wills itself spontaneously into action in defiance of the basic laws of cause and effect.

This is not the way the world around us works. Clocks don't magically intuit time, electrical appliances don't work spontaneously without being plugged in or having a battery supply, wars don't happen by chance, crops don't spring up on their own without planting, and high tides don't occur randomly. Contrary to the bedrock conviction we all hold that we somehow are different, there simply is no proof of such.

Strictly as a matter of logic, the case for free will requires heavy lifting. Some free will doubters attempt to split the difference. In philosophy circles their position is called *compatibilism* (compatible with free will *and* determinism). But their efforts seem little more than overly convoluted attempts to deny the obvious. In his book, *Free Will*, Sam Harris, a neuroscientist and philosopher, puts it this way: "Compatibilists have produced a vast literature in an effort to finesse the problem. More than in any other area of academic philosophy, the results resemble theology" (Harris, 2012). Nuclear physicist Alan Scott characterizes compatibilism as a failed effort to have it both ways (Scott, 2018). Reading over various versions of compatibilism, one senses a strong resistance to relinquish the idea of free will for fear it will open the door to hedonistic and unbridled antisocial behavior. Since in the absence of free will, no one could be held responsible, anarchy would be a certainty (Klemm, 2014). But as we shall see later, this is an unnecessary assumption as long as you retain *responsibility*.

(Blameless responsibility is a subject we will go into later in considerable detail.)

Despite the almost universal acceptance of free will, there is a long list of unflinching dissenters even before neurobehavioral science entered the picture. Philosopher Baruch Spinoza rejected the idea with a surprisingly modern explanation for why the conviction persists. "Men are mistaken," he asserted, "in thinking themselves free; their opinion is made up of consciousness of their own actions and ignorance of the causes by which they are determined" (Spinoza, 1992). The 19th century British biologist Thomas Huxley categorically denounced free will this way: "... the feeling we call volition is not the cause of a voluntary act, but the symbol of that state of the brain which is the immediate cause of that act" (Huxley, 1894/1911). In his satirical, science fiction novel, *Slaughterhouse-Five*, Kurt Vonnegut portrayed one of his other-planet characters trying to understand this strange idea. "If I hadn't spent so much time studying Earthlings... I wouldn't have any idea what was meant by free will," he explains. "I've visited 31 inhabited planets in the universe, and I have studied reports on 100 more. Only on earth is there any talk of free will" (Vonnegut, 2009). The discoverer of *relativity*, Albert Einstein, left no room for doubt as to where he stood on the question when he addressed a German Spinoza Society meeting in 1932: "... human beings in their thinking, feeling and acting are not free, but are as causally bound as the stars in their motions" (Einstein, 1932). Later Einstein was even more blunt in his book, *My Credo*. "I do not believe in freedom of the will," he asserted, explaining how the philosopher Schopenhauer had it right when he said: "Man can do what he wants, but he cannot will what he wills" (Einstein, 1932).

Where We Are Headed

This last *come down* — the questioning of the human ability to *freely* consider options, deliberate, choose and then act—informs the rest of this book, but we start in the next chapter—*Rule Breakers and Blame*—with a consideration of the confusion the free will assumption has produced with respect to how we apply blame and punishment. In our minds it's a straightforward proposition: assuming we truly are captains of our own fates, willfully choosing our way through life, if we opt to break rules we are *fully blameworthy*. This deeply ingrained human conviction provides the foundational justification for our approach to criminal justice. Chapter 2, *Getting Blame Right*, considers the difficulties encountered trying to assign blame *consistently and fairly*, the enduring practice of scapegoating, and the critical distinction between blame and responsibility. *Two Lives* (Chapter 3) graphically illustrates the complexity of traditional blame assignment by recounting the different fates of two cold-blooded murderers—Karla Faye Tucker and Nathan Leopold—and the criminal justice awkwardness of reform. In Chapter 4, *The Devil Made Me Do It and Other Defenses*, we take a look at the close-guarded *defenses* against blame, starting with insanity. Chapter 5, *The Scourge of Massive Incarceration* (some of it based on my own personal observations as a prison psychiatric consultant) provides a no-holds-barred account of the criminalizing hellholes used to inflict punishment for the sake of punishment. Chapter 6, *Intricate Dance: Genes and Experience (Environment)* drills down on the causes of human behavior with the main takeaway: when searching for answers to why we think, feel, and act, we need look no further than the combined effect of *genes and experience (environment)*. At Chapter 7, *The Astonishing Illusion*, we arrive at what we have been building toward from the start. As revealed by neurobehavioral science, the self as it turns out only has the appearance of being a choice-making action figure. In fact it is the chief protagonist in an illusionary after-the-fact summarizing narrative and in so being is beyond blame. *Deep Story Telling and the Self* (Chapter 8) further elaborates the illusionary self in its role as central character in the life stories that dominate human experience. Chapter 9, *Relocating Evil*, reconsiders our traditional view of evil as residing within the person and instead makes the case for relocating it outside the person in the actual causes and evil acts themselves. Chapter 10, *Blameless Justice*, lays out a radical *alternative approach* to criminal justice based on *blameless responsibility* absent punishment-for-the-sake-of punishment. Finally, Chapter 11, *Conclusion*, provides a summarizing overview of the illusionary self and the misguided approach to blame and punishment it inspires, the need for rethinking evil, and an alternative view of blameless criminal justice.

Although there's nothing overly technical about this book, it does explore a mind-bending concept or two. My promise is this: what you are about to read will be provocative, intriguing, and thought-provoking. Its main thesis—though seemingly outrageous at the start—by the end of the book will become a much more acceptable proposition; one that—whether you agree or not—will leave you thinking long after you finish. In the next chapter we turn to rule breakers and the blame we falsely assign them.

Chapter 1

Rule Breakers and Blame

Fixing blame is a delicate assignment.
Browning Ware

Based on the assumption that we all choose to do what we do, blaming is a deeply ingrained response to rule breaking. To the extent we blame, we cannot be empathetic. The idea that had we been in the exact same position, we would have done the same is excluded.

If someone knowingly and willingly breaks the rules, it's on him or her. They are to blame and deserving of punishment. The greater the transgression the more severe the punishment.

Blame is our default response to rule breaking as is trying to avoid blame when we are the object. One of the oldest methods is to try and shift the blame by putting it off on something or someone else. Favorite targets include a menagerie of sacrificial animals, devils and daemons, and of course other persons. But the list of blame objects is virtually unlimited. Even eating utensils have been singled out to take the fall!

Scapegoating

Blame is sometimes assigned in a highly ritualistic manner. Scapegoating is one of the oldest. The word itself comes from the ancient practice of sacrificing one goat to Yahweh on the Hebrew Day of Atonement and releasing another into the wilderness after a high priest has symbolically transferred to it the sins of the people (Campbell, 2011). In ancient Greece the Athenians reworked the scapegoat theme into their harvest festival of *Bouphonia*. In what was quite an elaborate production, oxen were herded into a sacred area where barley cakes rested on an altar. The first animal to go for the cakes had its throat slit after which the "murderers" threw down their weapons and fled. Later, as onlookers feasted on the slaughtered animal, a "trial" took place which consisted of a long litany of dramatic finger pointing. The women who had carried the water to wet the knives accused the knife sharpeners. In turn, the sharpeners impugned the butchers. Finally, the butchers, having no one else to blame, turned on the knives. In a great show of moral outrage, the trial was adjourned so the angry crowd could hurl the knives into the sea. With the community now fully absolved of all wrongdoing, a riotous celebration broke out which included new wrongdoing (Young, 1992).

In his monumental look at societies around the world (*The Golden Bough*) Sir James Frazer described numerous variations on the scapegoating theme, some benign, others horrific (Frazer, 1994). In Indonesia villagers built small boats and put them afloat to carry away sins. For a similar purpose, Himalayans stoned dogs to death. In Albania "sacred" slaves were imprisoned in the Temple of the Moon. Each year one was selected and, after being covered with sacred oils, chained and speared to death in an attempt to transfer all wrongs to the corpse.

If a disaster of some kind struck ancient Athens, at city expense one man and one woman were chosen from a group of human "outcasts" housed for just such an occasion. The two unlucky persons were then dressed in fine clothes and jewelry to be paraded in public while prayers petitioned the deities to transfer all the city's sins to them. Eventually, the macabre procession made its way out the city gates where the populace gathered to stone to death the two celebrants before stripping them of all their finery. In a variation, the Greeks of Asia Minor employed a standing supply of *sin atoners* composed of dwarves and persons with various physical deformities as the scapegoats.

One of the more bizarre examples of scapegoating involved putting *animals* on trial.

In Bale, England (1474) a rooster was tried for laying an egg. When the defense's claim that the rooster's perverted act was *involuntary* proved unpersuasive, the rooster was declared guilty of being a shape-shifting sorcerer with a predilection for birds. Together, the perverted rooster *and* the egg were burned at the stake. Other animals put on trial included weevils, caterpillars, snails, wolves, and locusts (Evans, 1987).

During the middle ages, "sin-eaters" were used at funerals to help the recently deceased progress from purgatory to heaven. Their role was to sit next to the corpse and consume huge quantities of food. In doing so, they were thought to be devouring the dead person's sins. (Keep in mind, as bizarre as all these examples seem, later we'll see how similar they are to modern acts of blaming, once the illusionary basis of blame is understood.)

A better known example of scapegoating is the New England version (1690s) where witch hunts were devised to ferret out persons in league with the devil. One of the most damning findings was any minor bodily imperfection. As a result, even on the flimsiest of evidence, women were blamed, pressured and tortured so that at trial many of them "voluntarily" described being raped by the devil. The New England version was only one of many witch hunts. Scholars estimate worldwide between the years 1450 and 1750 roughly 100,000 witchcraft trials resulted in 40-50,000 executions, the vast majority women, hanged, drawn and quartered, or burned at the stake as a presumed means of protecting others from evil (Worthen, 2017).

Behind all these various practices is the assumption of blame, but as you can see through the centuries administering blame has proven to be a messy, imprecise, and often deadly affair. Little has changed in the modern world. In fact some have portrayed our criminal justice system itself as just another— albeit, more sophisticated—form of scapegoating where select rule breakers—mostly drawn from marginal elements of the population—are overly blamed.

Blame in Religion and Art

Blame finds solid support in many religions. The Old Testament advocates "an eye for an eye"; the Koran provides detailed instructions for "honor killing." Blame is also well represented in the arts as a major driver of human behavior. *Hamlet, Othello,* and *The Merchant of Venice* are only some of Shakespeare's plays with blame front and center. Similarly, blame drives Verdi's opera, *Rigoletto,* and novels such as Dickens' *A Tale of Two Cities,* Hawthorne's *The Scarlet Letter,* and Dostoyevsky's *Crime and Punishment,* to name three of many, are filled with blame and vengeance similarly portrayed in the visual arts in works such as Caravaggio's painting, *Salome with the Head of John the Baptist.* Blame appears frequently in modern musical lyrics such as The Dixie Chicks' *Goodbye Earl,* Taylor Swift's *Bad Blood,* the musical *Chicago's Cell Block Tango,* and Justin Townes Earl's *Someone Will Pay.* And numerous modern plays and movies center on vengeful blame well illustrated by *Who's Afraid of Virginia Woolf?, Death of a Salesman, Mystic River, Dead Man Walking, The Hurricane,* and *Unforgiven.*

Blaming Hysterias

There are times when blame spikes as a *collective experience* in response to spurious wrongs. The witch trials, the Holocaust, and other periodic ethnic-targeted pogroms are all examples. More recently in the 90s a new epidemic of alleged evil appeared in this country when psychotherapy patients began to report abuse involving bizarre satanic sexual practices undertaken by family members or other persons in their communities, some well-known and influential. Many of these cases were reported by children. Before this rolling storm of what turned out to be false blame burned out, innocent people went to jail, sometimes for long periods.

In August 2017, Dan and Fran Keller received 3.4 million dollars from a Texas state fund for wrongful convictions (Boroff, 2017). In 1992, while running a well-established and respected licensed day-care center in their Austin home, the couple was charged with child sexual abuse and eventually convicted, primarily on the basis of unsubstantiated claims by children attending the center who insisted the Kellers had forced them to watch ritualistic sex practices and the dismembering of babies. At the trial's conclusion the Kellers were sentenced to 48 years in prison. *Twenty-one years later* their convictions were overturned after an appeals court discovered a testifying doctor had flat-out lied about finding physical evidence of abuse. Eventually this blame hysteria subsided but only after courts found certain therapists guilty of implanting false memories that gave rise to these preposterous claims (Wright, 1994).

Blame versus Responsibility

Commonly blame and responsibility are used interchangeably. In fact you sometimes sense some who cling to the idea of free will do so out of fear that to do otherwise would absolve everyone of being responsible. Later we'll find even scientists who have contributed to the case against free will reluctant to accept their findings for fear of the disastrous consequences of unleashing a "blame-free" world. They get into this bind by falsely equating blame and responsibility. The two are not the same.

You can have responsibility *without* blame. This is a critical distinction for the rest of our discussion. So, how are they different? Let's start with responsibility. It implies an obligation to assume consequences for a broken rule. Essential for social stability, it comes with human membership. In certain instances such as mental retardation or dementia, responsibility may pass to others or to the state; even so, *responsibility must be retained*. Without it social chaos would be inevitable. But full responsibility with consequences is fully compatible without the inclusion of blame and the punishment.

In 2004 two Princeton University psychologists, Joshua Greene and Jonathan Cohen, wrote a fascinating article ("For the Law, Neuroscience Changes Nothing and Everything") that anticipates several themes in this book. Predicting the coming assault of neurobehavioral science on free will and our current "retributive" approach to criminal justice, they outlined an alternative based on consequences flowing from responsibility, *not* blame (Greene, 2004). The *responsibility perspective* emphasizes liability for actions without implying culpability. (Think no-fault divorce.) Imagine a police unit steeped in this assumption. Although violent crimes would still require containment, they along with other police actions would be free of the anger and excess that sometimes seem justified by the belief the perpetrator *deserves* what he or she is getting. This is the point where many of my friends and colleagues roll their eyes and chuckle as they proceed to tell me how I have totally lost my mind. Bear with me.

Strict Liability

Blame has a central place in Western law. Typically the amount of blame assessed depends on the seriousness of the broken rule—ranging from being of little consequence to the taking of life—in combination with the perpetrator's alleged state of mind; willful, premeditated, cold-blooded and intentional being considered the most blameworthy. Underlying it all is the assumption that rules are broken by *choice*. But in more recent times, a variation has evolved. In both civil and criminal law, laws have appeared that while automatically conferring consequences when they are broken (often in the form of fines) imply no fault or blame. These laws are characterized as exhibiting *strict liability*. The rule breaker's mental state, willingness, intention, or motivation at the time of the offense is *irrelevant*. The only germane question is whether or not the person broke the law. If such is established, he or she is deemed guilty of the infraction and subject to a pre-established penalty *without fault*. Whereas blame emphasizes retribution in response to rule breaking, strict liability focuses strictly on evidence the law was broken and pre-prescribed consequences.

This is the perspective, for example, behind the no-fault issuance of parking tickets (Greene, 2004). Necessarily there are certain restrictions on the parking of cars as to where and for how long. When these restrictions are disregarded, responsibility falls to the registered car owner as an obligation of ownership. No questions are asked about why the car was parked illegally; no queries are made about intent or state of mind; and, for the most part, no excuses are accepted. Upon discovering the parking citation, the responsible owner may grumble about it. Perhaps it was haste in getting to an important class or appointment that caused him or her to overlook the "no-parking" sign; or, perhaps it was a friend who parked the car illegally. Whatever the reason, from a *responsibility perspective*, it is irrelevant. Fault or blame is not the issue. Only the obligation that goes with owning the car and committing the infraction matters. No need to probe for signs of malice, premeditation, or irresistible impulse. Even a claim of insanity is beside the point. The law doesn't care *why* the car was in the wrong place. A rule has been broken. There's a standard penalty. The responsible owner is obligated to pay a fine. If the owner pays, nothing more is required. He or she has been held responsible *without* blame. No need for retribution. No place for punishment. Important to note, this development in the legal field came without any questioning of traditional assumptions about free will and choice. Looking ahead, if those assumptions are questioned, the potential scope of strict liability becomes greatly expanded.

But what about more serious rule breaking? How would

the idea of strict liability work if applied in those cases? Of course, stated consequences would become more severe as the seriousness of the rule breaking increases. Conviction for a DWI, for example, might carry a fine as well as a prohibition on future driving. But not punishment per se. Containment might eventually be required but as a public health measure not punishment. Responsibility for driving a car safely implies remaining sober. A person who drives intoxicated breaches his responsibility and signals that more intrusive measures are required *as a matter of public safety*. As for extremely dangerous rule breakers, holding them responsible might well entail lifelong containment, again, not as punishment (and not in punishment settings) but as a means of public safety. Notice I used the word "containment," not mass incarceration. As we will see later, the shift in perspective from blame to responsibility has enormous implications for correctional methods.

So, we have these two contrasting views of rule breaking. The well-established *blame perspective*—informed by our subjective experience where the self seems to act on its own, free of external influences—assumes that when rules are broken, blame and varying degrees of punishment are appropriate if not essential. This is in contrast to the *responsibility perspective*. Here, there is no case to be made for blame or punishment; only responsibility and obligation in the form of clearly established consequences. The rule or law is broken and there are straightforward consequences, but no punishment.

In our current approach to rule breaking, blame and punishment are foundational assumptions. If they were put aside much of what goes on under the rubric "criminal justice" would lose its justification. Laws, courts, judges, juries, policies, settings, personnel, and practices—none of these would remain unchanged when redesigned based on the principle of *blameless responsibility*. Resistance to such changes likely would be fierce. (So many vested interests.) But if what we take up later

Chapter 2

Getting Blame Right

There's a saying that a criminal trial is a contest held in public in fancy dress to decide which side has the best lawyer.
P.D. James

The common impression is that in order to draw the death penalty, you have to be guilty as sin of a truly heinous crime. Wrong. You don't have to be accused of a particularly horrible crime to draw the death penalty; you don't even have to be guilty.
Molly Ivins

As poignantly illustrated by massive protests over police brutality, our official criminal justice system as a whole is replete with questionable practices. For starters, a shocking number of persons retained in our nation's jails are held there without conviction simply because they do not have the funds to post bail or bond while their cases are being processed. The Vera Institute of Justice Centering on Sentencing and Corrections estimates this injustice applies to 60% of all persons in jail at any one time. This despite the fact that three-quarters of these pretrial jail detainees are being held for nonviolent traffic, property, drug or public disorder offenses (Vera Institute of Justice, 2015).

Deeper into our criminal justice system the problems only get worse. *Official* blame carries the expectation that similar wrongs deserve similar punishment. Although seemingly straightforward, this guideline proves impossible to follow. Despite numerous judiciary formulas, glaring disparities in the assignment of blame persist. The same crime commonly elicits different punishments; and, as we shall see, it is not rare at all for perpetrators of more serious crimes to receive lesser punishment than those who commit minor offenses.

All attempts to eliminate this disparity have failed. For practical reasons, equivalency is out of the question. Consider the demands on a criminal justice system dedicated to imposing murder for murder, rape for rape, embezzlement for embezzlement, burglary for burglary, or assault for assault. But absent handing out equivalent wrongs, punishment is routinely dispensed inconsistently and unfairly.

Take for example *ultimate* blame and the punishment that goes with it, death. Currently, murder is *the* death penalty offense, but not just any kind of murder. It has to be *aggravated* by special circumstances such as multiple murder, murder committed during the commission of other crimes (such as kidnapping, robbery, arson or rape), "heinous" murder, murder by hire, murder of law enforcement officers or firefighters, and murder of crime witnesses. At first glance these aggravating conditions seem reasonable, but in reality what seems straightforward proves troubling.

"Special Circumstances"

For example, what makes a murder *heinous*? In part, it depends on the weapon. A single shot from a pistol likely will not qualify; scissors, maybe; a pitchfork, probably; a pickax or chain saw, definitely. Despite the same tragic outcome (a person's death), the kind of weapon determines that one killer is executed and another lives. A murder is also more likely to be judged "heinous" if it is bizarre or primitive in appearance. But why should the designation of an offense as *capital* depend on the weapon or the nature of the killing? A murdered person is a murdered person.

The majority of capital murders fall into the category of *felony murder* (murder committed during another felony). Here again, definition can be tricky. Take *kidnapping*. If you rely on movies and novels, you would assume kidnapping means abducting and hiding a person as a setup for ransom, but the actual legal definition is far more nuanced and complicated, ripe for inconsistent application. Consider what happened in Jasper, Texas, during the early morning hours of June 7, 1998. Three young white men offered a ride to James Byrd, Jr., a black man; and, then, not far down the road, stopped and chained him to the back of their pickup truck (Rosenblatt, 2013). Zigzagging from one side of the road to the other, they dragged this man's body for miles long enough for Byrd's head and limbs to fall off. The three killers were quickly taken into custody. Evidence against them seemed airtight.

But surprisingly right from the outset the prosecutor was concerned this horrendous murder might not qualify as capital murder. He understood that simply because the three young men were avowed white supremacists did not reach the threshold. (At the time there were no "hate crime" laws.) In searching for legal technicalities, the prosecutor discovered what he thought might be a loophole: if federal funds had been used in the construction of the road along which the victim was dragged, the crime might be "death qualified" under federal (not Texas) statutes. But this line of reasoning ended up leading nowhere. Eventually, the prosecution was only able to justify a charge of "capital murder" by combining the crime of first-degree murder with *kidnapping*.

As it turns out, the legal definition of kidnapping turns on the question of restraint. Any substantial interference with a person's movement qualifies. Still, in this case to charge kidnapping would depend on proving Byrd was *alive* at the time he was chained to the back of the truck; so, only when the medical examiner confirmed Byrd had died afterwards as he bounced along a country road did a charge of felony murder become possible. By itself, under existing capital law, this savage, racist killing did *not* qualify for ultimate blame and punishment.

When more closely studied, special circumstances read like a list hastily thrown together by a disinterested committee. Why restrict the death penalty to *aggravated* murder, anyway? Why would intentional murder of any kind not qualify? Why should ultimate blame be assigned for the murder of policemen and firemen but not for the murder of an EMS worker or teacher or nurse?

The definition of *felony murder* has become so convoluted even persons not directly participating in the killing can be sentenced to death. Hypothetically, the driver of a get-away car for a botched robbery and unplanned murder can get the death penalty. Contrast this with a person who commits malicious, premeditated murder but, because there is no aggravating factor, can't be put to death. The implication is clear: if you are disgruntled with your business partner enough to want him or her out of the way, then do the job yourself. Plan it, purchase the weapon (keep it simple), and carry out the murder in his or her home. If you get caught and have not deviated from this prescription, your life will be spared. But be careful, with the slightest deviation all bets are off. For example, if you commit the murder in your car, you run the risk of converting the crime into felony murder aggravated by kidnapping. Likewise, if you hire someone to do the killing for you, it's felony murder. Same crime, same weapon, same victim's death, but any of these variations can mean the difference between life and death. Such are the puzzling and irrational inconsistencies of capital blame and punishment.

Why not apply ultimate blame to *all* cases of murder? If a person is convicted of killing another person, he or she gets the death penalty. Period. A life for a life, plain, simple and fair. It seems straightforward enough, but like all attempts to clarify the death penalty, this proposal has serious problems. Few among us—including the staunchest supporters of the death penalty—are willing to treat all instances of human killing as capital offenses. To do so would lump together premeditated and sadistic murders with accidental killing, killing under-the-influence of drugs or alcohol, psychotic murder, and killing by a minor. Obviously, a more circumscribed definition is needed, but try coming up with it. Not so easy.

Perhaps a specific murderous crime could serve as a "gold standard" for *capital* murder. What about the Oklahoma City bombing? Timothy McVeigh carried out a carefully calculated terrorist act that killed 168 random victims (Russakoff, 1995). One editorial writer put McVeigh's crime in perspective: "In this one violent act, he took more lives than all 21 murderers combined who were executed in Texas the same year." Few would argue if there is to be a death penalty that McVeigh didn't deserve it. But if McVeigh's crime is at the top, a ten on the murder scale, how would lesser murders be rated? What kind of murder would merit a 7 or 5? And where would be the cut-off line for requiring death as punishment. Surely a developmentally disabled person who impulsively shoots a storeowner shouldn't receive the same punishment as Timothy McVeigh? But what should the punishment be. How about an older teenager, bombed out of his mind on drugs, who kills and robs his parents when they refuse him money for his next fix? Or, a cigarette company executive who knowingly and deceitfully oversees the manufacturing and selling of addictive and disease-causing products responsible for the deaths of thousands each year? No surprise, the application of capital blame and punishment is riddled with inconsistencies.

Glaring Disparities

Just down the hall from where O.J. Simpson stood trial for the deadly slashing of his former wife and Ron Goldman, her friend, Ernest Dwayne Jones was being accused of the rape and murder of his girlfriend's mother. Julia Miller had been found dead, two kitchen knives protruding from her body. With no eyewitnesses to the crime, the prosecution's case depended mainly on DNA evidence. The same lab used for Simpson did the analysis in the Jones's case. But this is where the similarities end.

In stark contrast to Simpson's high-profile defense team, Jones's representation consisted of a single public defender. In his case DNA evidence was presented in one day. Why? Because Jones could not afford the DNA experts needed to tear apart the prosecution's case. Jones's lawyer did not even cross-examine the prosecution's DNA expert, and he never raised the possibility of contaminated blood samples—so effective in Simpson's case. The guilt phase of Jones's trial was over in 12 days. He got death.

In contrast the Simpson trial turned out to be a slam dunk. Drawing on the expertise of a high-paid, jury selection expert, Simpson's lawyers used their preemptory challenges well to seat eight African-Americans—six women, two men (two-thirds of the jury)—in a city where blacks are one-ninth the population. Closely followed by the media and the public, the trial rolled on for nine months as Simpson's legal "dream team" tied the prosecution in knots ("If it [the glove] doesn't fit, you must acquit.") before O.J. walked out of court a free man (Bugliosi, 1996; Wilson, 1997).

Two men. Both commit brutal crimes. Both are tried for murders not too dissimilar with dramatically different outcomes. Why? O.J. Simpson was a folk hero. Articulate, charming and handsome, raised by a loving mother, college educated, and a multimillionaire. A sports legend and a media star. Contrast this with Ernest Dwayne Jones. Jones was developmentally disabled, raised by alcoholic parents. As a child, he was severely beaten and sexually abused. He had no constituency. No celebrity. No money. No charm. And poor legal representation. The outcome was never in question. Blame is a hard thing to get right.

In 1978 former San Francisco city supervisor, Dan White, disgruntled over not being allowed to retract his recent resignation, shot and killed Mayor George Moscone and Harvey Milk, the city's first openly gay supervisor (Lindsey, 1985). The crime was carefully planned. Armed with a *pre-loaded*, snubnosed Smith & Wesson revolver, White crawled through a city hall basement window to avoid metal detectors and walked up a back staircase to Moscone's office. He fired four shots, the last two from inches away, into Moscone's head.

After leaving the murder scene undetected through a rear door, White made his way to another part of the building, reloading his revolver as he walked. When he found Supervisor Milk, he asked if he had a moment to talk. When they reached Milk's office, White killed him, the last shot assassin-style behind the ear. Despite the obvious deliberate nature of these murders, he was convicted of *voluntary manslaughter* with a *maximum sentence of seven years and eight months.* (A major defense claim placed much of the blame on White's eating habits, the so-called "Twinkie" defense.) Six years later, in 1984, White was paroled. (Later he would take his own life.)

Compare Dan White's crime and punishment to the pathetic case of Mario Marquez. In describing Marquez's life, his lawyer portrayed it as "sadness from beginning to end." He was the tenth of 16 children born to a migrant farm worker. When he was a child, his father savagely whipped him repeatedly. Marquez dropped out of grade school. (Reportedly, his I. Q. was 65-70.) Abandoned to the streets at age 12, he took up a life of vagrancy and drugs. In 1984, in a burst of jealous intoxicated rage, he raped and strangled his ex-wife and her 14-year-old niece. Eleven years later Marquez was executed by lethal injection, one of six retarded inmates executed since 1990 by the state of Texas. The following year, then Governor of Texas Rick Perry vetoed a bill barring the execution of anyone mentally retarded, claiming there was no need since Texas had never executed such a person (Bonner, 2001).

A more recent example of the pervasive inconsistencies of

"criminal justice" transpired on July 14, 2020. After a 17-year hiatus, the U.S. Justice Department decided to resume federal executions. Forty-seven-year-old Daniel Lee had been on death row in Terre Haute, Indiana for 20 years after being sentenced to die for the murder of three persons, one a child, as part of a scheme to fund a white supremacist group. Lee denied his guilt to the end claiming exonerating DNA evidence had been suppressed. A final appeal to the U.S. Supreme Court was rejected by a 5 to 4 majority, but Justice Stephen Breyer, writing for the minority, pointed out the glaring inconsistency of the case. How, he asked, could it be called justice when a co-defendant in the case, considered the more culpable of the two, received a life sentence while Lee was condemned to death? Breyer's question goes unanswered.

The explosion of so-called *mitigating circumstances* as an offset to blame, though well meaning, further complicates its consistent assignment. Each new discovery of factors that influence human behavior brings another potential basis for mitigation. Abuse, bad genes, drug addiction, PMS, hypoglycemia, sleepwalking—the list continues to grow. To what extent should punishment be lessened by subjective claims that a person's willpower was altered by influences beyond his or her control? And are there certain crimes so heinous they override any consideration of mitigating circumstances? If so, where should the line be drawn? How terrible must the crime be? Mitigation has become a bottomless pit of ill-defined potential defenses (labeled by some as *abuse excuses*) against wrongdoing. To further complicate matters, despite these perplexing questions, the U.S. Supreme Court has ruled *all* mitigating factors must be considered.

Capital Punishment: Just or Capricious?

Nowhere are the inconsistencies of legal blame more obvious than in the use of death as punishment. The seemingly simple proposition of taking a person's life as ultimate punishment for taking life has proven fraught with ambiguity and human error. What about a person who sexually savages and murders 17 young men? Would the death penalty be a sure thing? Not necessarily. Jeffrey Dahmer, not content with sexual assault and murder, carved up the bodies of his victims, boiled the parts, and ate some (Kennedy, 2016). Afterwards, he kept some of his favorite portions for souvenirs before disposing of the rest. Dahmer claimed insanity, but the jury rejected this defense. Still, despite the savage and primitive nature of his crimes, he was found undeserving of the death penalty. Compare the punishments for O.J. Simpson, Ernest Dwayne Jones, Dan White, Mario Marquez, and Jeffrey Dahmer. O.J. walks (only to be tried and convicted later of another crime). Jones and Marquez are executed. Dahmer gets life, and White is paroled after six years in prison. The Menendez brothers, Erik and Lyle, murdered their parents for money yet still managed to escape execution (Davis, 1994). There's something seriously wrong with this picture. It's almost as though a random number generator is being used to hand out death sentences. Despite legal "guarantees" of consistent and proportionate use, capital punishment remains horribly capricious and biased in its application.

Over a period of 17 years, Theodore Kaczynski (better known as the Unabomber)—a mathematics professor turned revolutionary hermit—terrorized the country by dispatching carefully-crafted letter bombs from an isolated Montana cabin (Freeman, 2014). Before he was finally apprehended, he had murdered three people and injured 22 others. A coded journal from Kaczynski's remote cabin (deciphered by the FBI) provided a detailed record of his chilling reactions to news accounts of his murderous acts. After his first bomb killed Hugh Scrutton, the owner of a Sacramento computer store, his entry said: "Excellent. Humane way to eliminate some."

Thomas J. Mosser, a New Jersey public relations executive, was a Kaczynski victim. Upon opening a package delivered to his home in North Caldwell, Mosser was sliced to death by an exploding device that hurled a sheet of nails and razor blades across the room. Mainly because Kaczynski's brother (who had turned him into the FBI) argued strongly against the death penalty, his lawyers were able to plea bargain their way into a sentence of "four life terms plus 30 years in prison." This is how the man who maliciously injured or killed 25 persons dodged the death penalty.

Consider the following crime. An engaged couple argue over the man's sexual affair with another woman. As time passes, despite the man's denials, his fiancée grows more jealous until it threatens their engagement. Finally, together they come up with a solution. They will solve the problem by killing the 16-year-old romantic rival. A few days later the man calls and arranges to pick up the other woman around midnight. With his fiancée concealed in the trunk of the car, he drives out to a remote lake area. When they arrive, the fiancée climbs out and proceeds to bludgeon her rival in the head. Bleeding badly, the injured woman stumbles off into a nearby farm field before collapsing. The man follows and finishes her off with two shots to the head.

The crime goes unsolved, and the couple leave for college (but not before the murderer woman boastfully tells her friends what she and her fiancé have done). The boyfriend goes to the Air Force Academy; his fiancée to the Naval Academy. But, eventually, the two become prime suspects and confess to the crime. As they quickly fall into trying to incriminate one another, their stories change about the details of the killing (Meyer, 1998). True story. In separate trials, Diana Zamora and David Graham were tried and convicted of capital murder. Given the crime and where the trials took place—Texas—one would have assumed the death penalty for certain. Cold-blooded, premeditated, malicious murder, *aggravated* by kidnapping. What else but the death penalty? Sound reasoning, but as it turned out, incorrect.

The state chose *to forego the death penalty in both trials.* Diana Zamora's defense lawyer portrayed the killing as a "macho" act by Graham, designed to win back the affections of his girlfriend while David Graham's attorney characterized his client as a naïve, misguided lover covering up for his insanely jealous fiancée. Zamora and Graham were sentenced to life in prison with a minimum of 40 years. Why not the death penalty?

I put the question to an Assistant District Attorney in one of Texas' largest urban counties. He explained most likely it was the matter of the defendants having no previous criminal records as well as the outward appearance of upstanding character prior to the crime. Hard to imagine a similar outcome for poor and disadvantaged defendants of color.

It comes down to this. Despite mandated guidelines and procedures, the handing out of executions is remarkably subjective and inconsistent. In the three states with the largest death row populations (California, Texas, and Florida) less than 5% of murderers receive a death sentence; and, of these, many are never actually executed. Who gets the death penalty varies greatly from state to state. Compare, for example, New York and Texas. In 1995, New York re-instituted the death penalty. Three years passed before jury selection took place for the first capital murder trial under the new law. That same year, 21 states with death penalty statutes on the books failed to carry out a single execution (Death Penalty Information Center, 2017). In contrast, Texas powered ahead, leading the country in executions with 37 in 1997, half of the nation's total. No question about it, handing out the death penalty resembles Russian roulette far more than it does finely honed justice. If you decide to commit a murder, don't do it in Texas.

Innocents

The most disturbing aspect of using death as punishment is the possibility of getting it wrong. Without doubt there are those persons who have been executed by mistake. Reviewing 20th century murder convictions (1900-1991), two academic researchers, Radelet and Bedau, turned up 416 cases of false imprisonment and 23 wrongful executions (Radelet, 1998). Continuing their efforts, these researchers found on average 12 cases a year of erroneous death penalty sentencing.

But proving innocence and wrongful death, after the fact, is a tough assignment. Most efforts cease once a person is executed. Would it make much difference, anyway, knowing the *precise* number of wrongful executions? How many are too many? Some would say whatever the number, it is an unfortunate but unavoidable cost of doing the business of capital punishment. For others, one wrongful death is one too many.

In November 1998, the Northwestern University Law School convened a "National Conference on Wrongful Convictions and the Death Penalty." Since the 1976 reinstatement of the death penalty, 486 persons have been executed. The conference turned up 75 persons from 19 different states who had been wrongfully sentenced to die and lived to tell about it. That's roughly one person for every seven persons executed (Weinstein, 1998).

In his book, *Against the Death Penalty* (based on his dissent in the 2015 capital case, *Glossip v. Gross*), Supreme Court Justice Stephen Breyer references research showing approximately 4% of all persons sentenced to death are innocent (Breyer, 2016). It's enough to make him reject the death penalty as a legitimate punishment: "... I believe it highly likely," he says, "that the death penalty violates the Eight Amendment [cruel and unusual punishment]."

The Supremes and Death

The death penalty remains a badly-tattered piece of judicial work. Having finally struck it down in 1972 for being biased and inconsistently applied, only four years later the U.S. Supreme Court *re-instituted* it in *Gregg v. Georgia*. Its reversal was based on little more than paper promises by states of anticipated changes without any method of insuring they were actually carried out. Even concerning a matter as important as death as punishment, the Court found no reason to guarantee that stated correctives for making the process constitutional were real.

This is why for years afterwards Justice Thurgood Marshall conducted his own monitoring of death penalty appeals. Adamantly opposed to capital punishment, he viewed *Gregg* as a provisional decision, awaiting proof of results. Eventually, he was forced to conclude what he had suspected for some time: the death penalty remained both arbitrary and discriminatory. States continued to spare the lives of those who committed some of the "worst" crimes while executing others for lesser offenses with a persistent bias against "marginal" segments of the population.

Engraved in the stone façade above the main entrance to the United States Supreme Court building is the quotation: "EQUAL JUSTICE UNDER LAW." It is a worthy mission statement for the highest court in the land, but the inequities of capital punishment stand in obvious contradiction. Overseeing the rules for assigning ultimate blame and punishment has been a nightmare for the Court. Just prior to stepping down from the U.S. Supreme Court, Justice Harry Blackmun (appointed by Richard Nixon as a conservative jurist) expressed his change of opinion regarding the death penalty: "Rather than continue to coddle the Court's delusion that the desired level of fairness had been achieved and the need for regulation eviscerated, I feel morally and intellectually obligated simply to concede that the death penalty experiment has failed. It is virtually self-evident to me now that no combination of procedural rules or substantive regulations ever can save the death penalty from its inherent constitutional deficiencies... The problem is that the inevitability of factual, legal, and moral error gives us a system that we know must wrongly kill defendants, a system that fails to deliver the fair, consistent, and reliable sentences of death required by the Constitution" (*Callins v. Collins*, 1994).

Surely, from time to time, out of public view, the Justices discuss among themselves the Court's convoluted split-decisions and reversals regarding capital punishment. But despite this unenviable record, likely there is great resistance to admitting failure. To do so would imply the need for yet another full-scale review, and—given the Court's past performance—it's hard to imagine how such a thought would elicit much enthusiasm among the Justices, conservative and liberal alike, who now find themselves scrambling to limit appeals in an effort to speed up the frustratingly slow *and* inequitable workings of *capital punishment*. (After a 17 year hiatus, on June 29, 2020 the U.S. Supreme Court cleared the way for a resumption of federal executions.)

Adversarial "Justice"

Legal blame is made official in courts of law with adversarial rules of justice. Unfortunately, this transforms an alleged search for truth into a dramatic play where each side struggles to convince judge and jury of its preferred version of what happened, true or not. The goal of both is to create a distorted picture; in the one case, an exaggerated claim of guilt; in the other, of innocence. Facts have a hard time standing in the way. This bending of truth is legal and expected; standard operating and accepted procedure. In one review of 62 cases of wrongful conviction, researchers identified suppression of exculpatory evidence by the prosecution in 43% of cases (Radelet, 1992). Such are the ways of adversarial "justice."

F. Lee Bailey, the famed defense lawyer and a member of O.J. Simpson's legal defense team, once commented: "The worst client you can get is the guy on a dream cloud who says, 'I'm innocent so I'm sure to get off.' That attitude shows a lack of understanding of what a modern criminal trial is—a battle of gladiator-lawyers" (Bailey, 1972). Harvard law professor, Alan Dershowitz, puts it this way: There is "almost no correlation between the guilt or innocence of my clients and whether they served time or got off" (Dershowitz, 1983). Even in capital murder cases, the prosecutor has only to establish the appearance of guilt beyond reasonable doubt. Demonstrating the defendant's actual guilt is not the prosecution's burden. Any defendant who fails to appreciate this crucial distinction does so at his or her own peril. It's simply the way the adversarial "game" is played. This being the case, it is essential all defendants facing a possible death sentence secure the best legal representation possible, but for most, this is a total pipe dream.

Sad State of Blame Defense

Given that the majority of capital defendants are dirt poor, how do they retain competent legal representation? In all but a few high-profile cases, they don't. It is not that they are entirely without a lawyer (although some are). The Sixth Amendment guarantees "the assistance of counsel for his defense," and in *Gideon v. Wainwright* (1963) the Court nationalized the right to counsel in *all* criminal cases. But, as with so many things, saying it is one thing; doing it, another. If anyone needs *adequate* counsel, it's the person charged with a capital crime. But adequate counsel takes money. Lots of money.

Consider the case of George McFarland, tried and convicted for the killing of a Houston, Texas convenience store owner (Weinstein, 2000). The trial moved along like a bullet train. Opening statements and a guilty verdict arrived at in two days. A death sentence two days later. The entire case, including lunches and occasional recesses, consumed less than sixteen hours. *Houston Chronicle* reporter, John Makeig, described McFarland's state-provided counsel: "Seated beside his client... defense attorney John Benn spent much of the Thursday afternoon trial in apparent deep sleep. His mouth kept falling open and his head lolled back on his shoulders, and then he awakened just long enough to catch himself and sit upright. Then it happened again.

And again. And again. Every time he opened his eyes, a different prosecution witness was on the stand describing another aspect of the arrest of George McFarland in the robbery-killing of grocer Kenneth Kwan." When the Judge finally called a recess, one reporter asked Benn if he had really fallen asleep during the trial. The 72-year-old long-time (some would say too long) Houston lawyer shrugged and said: "It's boring."

Television viewers who followed the Simpson trial might assume state-of-the-art forensic tools such as DNA testing, high-powered investigators, and jury selection consultants are available to all murder defendants. No way. In most death penalty trials, at best, provisions for the defense are sparse without any guarantee of minimal legal competency. In the adversarial legal area, without necessary resources, a capital defendant resembles a gladiator without armor or weapon, which puts him or her at a distinct life-threatening disadvantage.

The Supreme Court's 1963 ruling failed to get into details. It did not specify a minimal amount of legal competency or remuneration, and it did nothing to assure funds for non-lawyer expenses such as DNA analysis which can be critical and quite expensive. Similarly, a second ruling in 1985 (*Ake v. Oklahoma*) failed to address these same issues. So, the matter was left up to trial judges who routinely are strapped for court funds. In essence this made the Court's ruling high-sounding but bogus.

There's no polite way to say this: the Supreme Court's guarantee of counsel to capital murder defendants for the most part is an empty promise. In the vast majority of cases, legal representation is little more than token. Correcting this problem would require resources far beyond what is provided. Absent the Court demanding delivery on its guarantees, fairness in capital murder cases remains mostly a mirage.

Strange Kind of Truth

Impartiality is *not* the objective of jury selection. Both the defense and the prosecution strive to create jury partiality. Jurors are human, they bring biases, conscious and unconscious, to their deliberative task. A jury that "leans" one way or the other, even before the case begins, can easily overcome the facts. This is why in high-stakes civil and criminal cases—where money is no problem—great attention is given to jury selection with considerable payment going to experts for help detecting and making use of juror bias. Wearing his novelist hat, Professor Dershowitz has a jury consultant describe ideal jurors to his client attorney: *"Older women,"* the consultant says, *"lots of them… with grandchildren. Stable families. No screwing around. No divorce. Kids who got married young. Miami Beach-in-the-winter types. Snowbirds. Italians, Irish, Jews, Greeks, maybe even some WASPs. No black women. No young women, regardless of race. And absolutely no well-educated or well-read people. Not dunces. Just not geniuses. And boring lives"* (Dershowitz, 1999).

Lies-for-lighter-sentences often find their way into criminal justice proceedings as does misleading "expert" testimony. Imagine jurors on a capital murder case being told by a state-appointed psychiatrist that if permitted to live the defendant would be *100% certain* to kill again. Texas psychiatrist, Dr. James Grigson, better known as "Dr. Death" repeatedly made this prediction, sometimes without ever having interviewed the defendant (Tolson, 2004). For substantial fees, Grigson traveled the state testifying in capital cases. The prosecutor would ask: "Can you tell whether or not, in your opinion, having killed in the past, he is likely to kill in the future, given the opportunity?" Dr. Grigson would reply with his standard answer: "He absolutely will… No matter where he is, he will kill again." Shockingly, jurors found Grigson extremely convincing, handing out death sentences in 93% of the cases (115 out of 124) in which he testified.

In *Barefoot v. Estelle* (1983), the Supreme Court upheld the testimony of two psychiatrists who readily admitted pronouncing a defendant a "continuing threat to society" without having examined the man. In a bizarre decision, the Court concluded psychiatric predictions of future dangerousness were not always wrong; just wrong "most of the time." Despite opposition from organized medicine, some states still make use of these so-called "killer shrinks."

In the final analysis, adversarial justice is a power game not a search for truth; and far more often than not, capital defendants have much less power or financial wherewithal than the prosecution. Although this inequity may be more acceptable in other areas of civil and criminal law, where human life is at stake, it seems straight-out wrong. Another product of blame-based justice.

Deal Making

The question of capital punishment does not always make it to a jury. There is a commonly used shortcut, known as *plea bargaining*. It is a way of arriving at guilt and sentencing through negotiation. In this country, plea bargaining is out of control. It has evolved into a way of "moving cases off the docket" to keep grossly overloaded courts from total chaos. But due to the power differential involved, it forces many innocent parties into accepting imprisonment rather than taking a chance of losing in court and facing a much more severe sentence. Unfortunately, there are no official rules governing these "off-line" deliberations. As a result, plea bargaining often deteriorates into a method by which prosecutors guarantee they come away with something. If the charge will be difficult to prove, no problem, settle for a lesser charge; or, if more than one person is charged, cut a deal with one of the co-defendants based on his or her promise to testify against the other. Unfortunately, plea bargaining, like adversarial justice, carries no commitment to the truth. The sole objective is to get a conviction... regardless. Put a win in your column. In her book, *Charged*, Emily Bazelon makes the case for how prosecutors, more than judges and juries, have come to dominate the outcome of criminal cases through their control over plea bargaining, selection of charges, and determination of bail, making what is advertised as a fair contest between competing views far more one-sided than what is commonly acknowledged (Bazelon, 2019).

It's rather easy for prosecutors to find witnesses willing to perjure themselves. One of the most likely "pickup" spots is jail or prison. So-called "jailhouse confessions" elicited through "snitch testimony" from cellmates are commonplace. The offer of shorter sentences or monetary reward does wonders to generate "incriminating" evidence. Even though the law requires prosecutors inform the defense, these deals often go unacknowledged. The fact that an informer can face future perjury charges serves as a compelling deterrent to these "agreements" ever becoming public knowledge. Thanks to plea bargaining, it is not uncommon for parties to the same murder to draw different sentences.

With courts and prosecutors badly overworked, plea bargaining becomes a go-to strategy. When you don't have the goods on someone, particularly a capital defendant, act as if you do and open the door to a trade. Using the threat of death, get a false admission of guilt for a lesser crime. For testimony essential to closing a case against a co-defendant, ignore who is actually guilty; offer a lesser sentence to the most willing participant. Whatever you do, get the case closed and move on. The considerable inequities of plea bargaining remain largely under the public radar screen.

Beyond Reasonable Doubt

With respect to court evidence, there are different *burdens of proof*. A civil trial dealing with private matters has a lower threshold than a criminal trial. Supposedly, this reflects the relative seriousness of the rule breaking. Bad to lose your money or your name, but not nearly as bad as being imprisoned or losing your life. The American public went to school on these differences as it followed O.J. Simpson through his two trials; the first, criminal, the second, civil. In the first trial (criminal), with a more exacting standard, Simpson was found *not* guilty, but in the civil case, guilty.

In civil cases the convicting standard of proof is "by a mere preponderance of evidence." The jury must conclude more evidence points to the defendant's guilt than away from it. In criminal proceedings, proof of guilt must be "beyond a reasonable doubt." It sounds definitive, but what does this actually mean? Surprisingly, there is no official definition.

When it comes to life or death decisions, you would expect the most rigorous standards. After all, the stakes are as high as they get. But the facts are otherwise. Over a 24-year period, 87 persons on death row had convictions overturned by exonerating evidence. This error rate—*one out of seven death penalty sentences*—suggests the term "beyond a reasonable doubt" is far more subjective than it sounds (Scheck, 2000). A National Science Foundation report found this standard a persistent problem for death penalty jurors. All one can dependably infer about "beyond a reasonable doubt" is that it means less than *all* doubt. How much less is up for grabs.

With the advent of DNA fingerprinting, the disparities in guilt decisions have become even more glaring. The *Innocence Project* has led the way applying DNA analysis to what turn out to be wrongful decisions. Still relatively early in its efforts, the Project has exonerated 350 persons (20 of them on death row) who on average had been falsely imprisoned for 14 years (Sapolsky, 2017).

In a modern world informed by a more nuanced sense of justice, blame assignment faces an ever-growing challenge. The first priority should be to get it right; to make certain it is handed out on a fair and consistent basis. But, as we have seen, decisions about blame and punishment appear more lottery-like than matters of justice. Different degrees of punishment for similar crimes is the rule not the exception. Establishing *someone's* blame becomes the only thing that matters. But the difficulty achieving fair and just blame assignment goes much deeper than the mechanics. As we shall see, the more fundamental problem is the idea of blame itself and the misguided punishment that flows from it.

A Different Perspective

Following years of experience as a London magistrate, the British judicial reformer, Baroness Wootton of Abinger, addressed the "injustice" of justice by proposing doing away with *degrees of guilt* altogether (Wootton, 1964). A noted sociologist and criminologist, Wootton insisted fair assignment of blame was futile. Criminal trials should be divided into two phases: a first for determining guilt or innocence and a second for deciding what should be done. For Wootton, speculations about states of mind at the time of a crime had no place in judicial deliberations. The singular goal should be to protect the public and, *if possible*, to rehabilitate the criminal; not to punish.

Inevitably, ahead-of-their-time proposals like Wootton's have met with sharp rejections, portrayed as little more than enablers of lawlessness. Obviously, people are to blame for what they do and deserving of punishment. Without blame people would do whatever they want with complete immunity and resulting social chaos. Later we'll consider Wootton's dismissed perspective in light of new neurobehavioral science findings that reveal the illusionary nature of blame itself.

In the next chapter we consider the vagaries of blame and punishment as illustrated in the lives of two persons both of whom committed horrendous crimes.

Chapter 3

Two Lives

We may have done a terrible thing, but we're humans. We change.
Karla Faye Tucker

The one thing I was known for was being a kidnapper and murderer. That was all right, but I was more than that. I was a human being too.
Nathan Leopold

Unlike times past, modern executions routinely occur years after the crime. Time enough for some murderers to undergo genuine change, transformed from perpetrators of horrific crimes into remorseful persons with remarkably changed views of life. By all appearances such individuals are no longer threats to society. For some it's simply a matter of growing up and leaving behind a troubled early life, one frequently marked by abuse and drug involvement. Others make their way to improved character through self-education or spiritual enlightenment. Whatever the path, some murderers truly are "reborn." Far removed from earlier acts of impulsive rage, intoxication, or sheer stupidity, they go on to become model prisoners. In fact, statistically, one of the safest groups of persons released from prison are murderers (Scheiber, 2010).

Even so, some reformed murderers are eventually executed the same as those who remain incorrigible, unrepentant, and unremorseful. While legally defensible, the execution of the truly rehabilitated remains an awkward and troubling aspect of capital punishment.

Born in different times, Karla Faye Tucker and Nathan Leopold were about as different as any two people could be with one major exception: early in their adult lives, they committed horrific murders. But once in prison they both underwent remarkable changes. Intertwined through the rest of this chapter, their stories well illustrate the difficulties of assigning blame and punishment appropriately.

Wired

For the writer Beverly Lowry it was a way of working through her own tragic loss of a son. After getting to know Karla Faye Tucker from visiting her in prison, she wrote a wrenching story of depravity and redemption in her book, *Crossed Over: A Murder, A Memoir* (Lowry, 1992).

Although it was only early summer 1983, already the Houston heat and humidity were stifling. After a three-day weekend of nonstop drugging, 23-year-old Karla Faye, accompanied by her live-in boyfriend Danny Garrett and his friend Jimmy Liebrant, drove over to Jerry Lynn Dean's Wind Tree apartment. They had been shooting crystal, drinking tequila, and taking an assortment of other drugs. Karla Faye had not slept for three days. "We were very wired... looking for something to do," she later tried to explain. "We went there to case the place..." The plan was to come back in a few days and steal a motorcycle.

Karla Faye had known Dean for some time. There was no love lost between them. The bad feelings started when Dean was dating Shawn, Karla Faye's former roommate. On one occasion when he parked his oil-dripping Harley on the living room rug, Karla Faye shoved him out of the house. The two got into it again when Dean and Shawn moved in together. Much to Dean's disapproval one weekend Karla Faye and Shawn went off to New Orleans with a rock-and-roll band. He blamed Karla Faye and took out his anger by cutting up some of her old photographs, including one of her deceased mother. When she discovered what he had done, Karla Faye punched Dean in the face and broke his glasses so that he had to go to an emergency room to have glass removed from his eye. Shortly afterwards, on the occasion of Shawn's dumping Dean, the two women stole $460 from his ATM account. Enraged, Dean put out a contract on Karla Faye for $300 to have her face burned with acid. When told about it, Karla Faye reportedly laughed it off claiming he didn't have the guts.

By any standard, Karla Faye's was a tough life growing up. At age eight she was already using drugs regularly—any kind she could get her hands on. She started shooting heroin at age ten, and the next year her mother, "Mama Carolyn," introduced her to prostitution. A few years later, when her mother's old boyfriend came to visit, Mama Carolyn off-handedly revealed to Karla Faye that her "father" was not her real father. It was this man!

Teaming up with her sister, Kari, Karla Faye ran away from home. The two of them started making trips to west Texas where they sold sex from a motel room in Midland. The money was good, and both women liked the free time. They referred to themselves as "call girls" since each had her own list of regular calling clients.

During one of these trips, Kari stayed behind. Over the weekend, she and Karla Faye's boyfriend Danny Garrett got loaded and had sex. When Karla Faye returned, Garrett confessed, and Kari made it worse by reporting how much she had enjoyed herself. Karla Faye shut down. This episode sealed it for her. She was on her own. Her wee-hours-of-the-morning drive to Dean's apartment happened a few weeks later.

The Perfect Murder

Nathan (Babe) Leopold's early life was a stark contrast to his crime. Born into wealth and privilege, at age eighteen in the spring of 1924 he was already in his second year of law school at the University of Chicago, set to transfer to Harvard the following fall (McRae, 2010; Baatz, 2008; Knappman, 1995). Members of two wealthy families, he and Richard (Dickie) Loeb had grown up in a fashionable section of Chicago. Both were notably intelligent. Leopold had an I.Q. of 210 and spoke eleven languages. Loeb was the youngest person ever to graduate from the University of Michigan. But the two men were strikingly different in physical appearance and personality. Loeb was handsome, tall and athletic, and despite a cold and haughty interior, came across as charming and easygoing, the life of the party. In contrast, Leopold was shorter and unsure of himself socially. Friends did not come easy which, in part, may account for his intense affection for Loeb. The two young men became involved in a sexual relationship more to Leopold's liking but encouraged by Loeb in exchange for support of his obsession with plotting outlandish crimes.

For several years Loeb had entertained the fantasy of becoming a master criminal. The two men often talked of crime and trained for bigger things by pursuing small-scale shoplifting, stealing cars, setting fires and randomly smashing windows without ever being caught or even suspected. On one occasion, fully outfitted with masks and guns, they robbed a fraternity house and came away with $74, a camera, a typewriter, a watch, and a fountain pen. Despite great apprehension, Leopold seemed willing to do almost anything to preserve the friendship.

Eventually, the two became caught up in Nietzschean philosophy. Filled with twisted views of the *superman*, they talked of pulling off the *perfect* crime. Something big. Eventually, they settled on murder. They went through a number of possibilities and even decided where to rent a car for the purpose of picking up a victim. Their idea was to set up the crime so they could jointly kill their victim, but when it actually happened, this was not to be (Baatz, 2008).

Late one afternoon they decided to go for it. Cruising the neighborhood where they both lived and still uncertain of their victim, they came across Bobbie Franks, age 14, dressed in a tan jacket and matching knee trousers, walking home from his Harvard Preparatory School. Franks sometimes played tennis at the Loeb estate. (In fact he was Loeb's cousin.) Leopold did a U-turn and pulled up to the curb. He asked if the boy wanted a ride home. At first Franks resisted, insisting he was already almost home, but he finally gave in. Loeb got out of the car and crawled into the back seat, and Franks slid into the front seat next to Leopold.

They had driven only a few blocks when Loeb grabbed Franks from behind and smashed the boy's head several times with the wooden end of a chisel. Jamming a gag down his throat, Loeb pulled Franks into the back and taped his mouth shut to stop his moaning. The two "brains" then drove leisurely through the south side of Chicago to a marshland near Lake Michigan close to the Indiana state line.

Somewhere on route, Bobbie Franks bled to death. Leopold stopped the car near a reservoir where he often bird watched. He and Loeb decided to have a couple of hot dogs and a root beer at a convenience store before finishing the job. When it was finally dark, they dragged Frank's body, now cold and stiff, from the car, removed his clothes, and poured hydrochloric acid over his face, presumably to make identification impossible. They carried the body 200 yards to a drainage culvert, stuffed it inside, and then drove back to Loeb's home where they burned Franks' bloody clothes in a basement furnace. Fatefully, Leopold failed to notice his glasses were missing having fallen from his pocket back at the culvert.

Pickaxed to Death

Karla Faye remembers getting out of the '77 Ranchero and walking up to Dean's apartment with Garrett. All the lights were off. When they reached the door, she opened it with a spare set of keys she had from the past. Later she reflected:

"I don't think we decided anything." Conscious decision or not, the plan was changing. While Liebrant stood watch, she followed Garrett inside.

It was a small 723 square foot apartment, completely dark. As the two groped their way through the living room and kitchen, they heard Dean in the bedroom and found him sitting up, naked on a floor mattress. Recognizing Karla Faye he called her name: "Karla, we can work it out," he shouted. (Presumably he was referring to the $460 taken from his ATM.) She cut him off and yelled back: "Move and you're dead, motherfucker," before sitting down hard on his stomach. Dean grabbed her arms and they rolled onto the floor until Garrett walked over and hit him in the head with a hammer he had picked up next to the door. Leaving Dean's body sprawled face down, Garrett took a "break" and walked out.

The gurgling sounds coming from Dean's chest irritated Karla Faye. She recalled how she kicked him to make them stop and then fell into a rage when the sounds persisted. She grabbed a pickax (Dean was a cable-television installer) and started hacking with abandon. Later she would boast to Garrett's brother, "Doug, I come with every stroke." It was a confession she would never overcome. (During her trial, a joke circulated through the Houston courthouse: the women's magazine, *Cosmopolitan*, was hot to get an interview. Seems the editors had heard Houston women had a new way of "getting off.") Despite Karla Faye's ax work the gurgling sounds persisted until Garrett returned. He rolled the body over and finished the job by splitting open Dean's chest. The gurgling stopped, and Garrett went back to moving motorcycle parts into the back of his Ranchero.

Only then did Karla Faye notice something shaking underneath the covers. As it turned out, Deborah Thornton, a 32-year-old bookkeeper and mother of a child, had met Dean the previous afternoon at a swimming party at her apartment complex. When the party ended, she arranged for child care, packed some things and drove over to Dean's apartment where, tragically, she arrived in the wrong place at the wrong time. As the first stroke of the ax grazed Thornton's shoulder, she lunged from under the covers, surely aware she was fighting for her life. When Garrett returned he found the two women struggling on the floor. Karla Faye extricated herself and left the room. By the time she returned, Thornton was sitting on the floor, her hands clutching the head of the pickax, now buried in her left shoulder, pleading for it to be over. Obligingly, Garrett kicked her in the head knocking her on her back and then axed her to death. Finished, Karla Faye and Garrett walked out together only to find that Jimmy Liebrant had already driven away.

Missing Glasses

When he failed to return home from school, Bobby Franks' family called the police but no real leads materialized until the following morning. About 10:30 a.m. the Franks' telephone rang. Mrs. Franks answered. (Just before the call came in, the police had told the Franks they had found a body resembling their son's description. Workmen had come across it accidentally.) The voice on the phone said the Franks' boy had been kidnapped and to expect a special delivery letter the following morning. It would contain instructions for putting $10,000 in old bills of certain denominations in a cigar box. The Franks were instructed to await another call telling them where to drop the money, but due to confused instructions this never happened.

Outwardly, Leopold and Loeb showed intense public interest in the case even as they talked to local journalists about the matter, and had it not been for Leopold's horned-rimmed glasses found at the crime scene, the two men might well have gotten away with their "perfect crime." As it was, the police quickly traced the glasses to a Chicago optometrist. He had written only three prescriptions for that type of eyeglass. The trail quickly led to Leopold, and after intensive questioning, he and Loeb confessed and began to angrily accuse one another.

"Well, Hell Yes"

Tucker and Garrett might also have never been caught if only they had refrained from bragging about what they had done. It was a few days after the killings when Karla Faye started boasting about how the murders proved she was ready to "play with the big boys." A month later Garrett's brother and his new girlfriend, Karla Faye's sister, Kari, tipped off the police, and Tucker and Garrett were taken into custody.

According to Charley Davidson, a former Houston prosecutor, at her trial Karla Faye Tucker looked every bit the part of a vicious murderer: "The embodiment of evil"... with "dark, lifeless eyes, like... a shark." The District Attorney decided to try the two of them separately with Karla Faye going first for the murder of Jerry Lynn Dean, and then Garrett for Deborah Thornton's murder. That way if things didn't work out to the prosecutor's satisfaction the crimes could be switched and the two tried again without double jeopardy. Karla Faye started off with a private attorney she didn't like and couldn't pay for. It wasn't long before her case was turned over to two public defenders.

Prosecutor Joe Magliolo hammered the theme of cold-blooded viciousness. Throughout the trial, whenever possible, he worked in the phrase, "hacked to death." Karla Faye's public defenders proved to be of little use. When the guilt-determination phase was drawing to an end, one of them rose to tell the jury he *agreed* with the prosecution. They should find his client guilty. It would be "an injustice for you to arrive at any other verdict," he asserted. As for her extensive use of drugs in the days leading up to the murders, he commented: "... if she went there so wired she didn't know what she was doing, she's still guilty of capital murder." The defense team presented no witnesses, and after 70 minutes of deliberation, the jury returned a guilty verdict.

The penalty phase did not go much better for Karla Faye. Prosecutor Magliolo persuaded Tucker's sister and brother-in-law to testify to her violent tendencies: fist fights with men and women, her compulsive workouts on the "speed bag," and Dean's pummeling before the murders. Both claimed they feared for their lives while living in the same house with her and described taking turns staying awake while the other slept. The defense called a psychiatrist retained with the government-limited $400 for a complete psychiatric evaluation. (This limit did not apply to the prosecution.) His testimony about Karla Faye's drug use and the possibilities of drug rehabilitation did little in the way of mitigation, and her lawyer's request for mercy from the jury was resoundingly countered by a question from the prosecutor. "Does Karla Faye Tucker deserve the death penalty? I'll let her answer that question," he concluded, as he punched the play button on a tape cassette player: "Well, hell yes," boomed Karla Faye's drug-husky, smoke-laced voice. The jury agreed.

Darrow's Defense

In its time, the 1924 Leopold-Loeb trial was as closely followed as the O.J. Simpson case. It was deemed "the crime of the century" (Higdon, 1999). The two wealthy Chicago families hired Clarence Darrow, aging but still the country's most famous defense lawyer, the man who would later defend John Scopes, a high school teacher of evolution, at the famous Monkey Trial in 1925. Darrow had represented 102 capital murder clients. All but one of his clients had escaped the death penalty. Part of Darrow's effectiveness stemmed from his own passionate opposition to capital punishment (Darrow, 2017). Long before the neurobehavioral science discoveries we will consider later, he believed no person was wicked by choice. It was always the result of heredity, environment, and chance elements over which the person had no control. Blame was misdirected. As for prisons, they brutalized these victims of circumstance rather than trying to humanely reform them (Darrow, 1996).

Once Darrow had apprised himself fully of the case, he rejected an anticipated plea of "not guilty by reason of insanity," intuiting it would be futile to try and convince a jury that these two extraordinarily bright men were crazy. Instead, at the last minute, he changed the plea to "guilty." With this change, the state lost the opportunity to try the defendants twice, once for murder and once for kidnapping, both of which at the time were capital crimes. Of even greater importance, the "guilty" plea avoided a jury. The life-or-death decision for these two men would be made by a single judge. It was the way Darrow wanted it. Face to face he would try to persuade the one man responsible for the verdict.

For two months, the case unfolded to a packed courtroom. In a lengthy and now legendary twelve-hour summation (delivered over three days), Darrow—one-on-one with the judge but in full view of onlookers—argued for the lives of Nathan Leopold and Richard Loeb. He pounded away at their young ages. No person in Illinois history under the age of 23 who pleaded guilty had ever received a death sentence. He countered the strong public impression that they would buy their way out. Confronting the issue head on, he reminded the judge that the death penalty was the same injustice whether a person had too little or too much money.

Making full use of findings from extensive psychological and physical tests (summarized in a 300-page report to the court) conducted on Leopold and Loeb, the aging legal warrior reserved his strongest attack against blaming the two boys for acts beyond their control (Epstein, 2018). He portrayed the judge as standing "between the future and the past," insisting that if he found these two young men responsible but blameless, the future was on his side. "Why did they kill little Bobby Franks?" he asked. "They killed him because they were made that way. Because somewhere in the infinite processes that make the boy a man something slipped." He characterized Leopold and Loeb as "unfortunate lads" hated and despised by "a community calling for their blood." Reportedly, by the time Darrow finished, it was clear most of the persons filling the packed steamy courtroom had been moved by what he said.

But at least one person in the courtroom remained unmoved. Prosecutor Robert Crowe followed Darrow with a scathing indictment, referring to the defendants as "cowardly perverts," "snakes," "spoiled smart-alecks," and "mad dogs." He asked the judge to "execute justice" and sentence the two men to die. It was the only way justice could be done. Lives for a life taken.

It took Judge Caverly three weeks to render his decision. At the sentencing he described how the murder itself was a "crime of singular atrocity;" but, acknowledging Darrow's impact on his decision, he spared the defendants' lives, sentencing them to "life plus ninety-nine years." Escaping the death penalty, Leopold and Loeb were immediately imprisoned in the Joliet penitentiary.

Changed Life

Early in Karla Faye's incarceration, she underwent a dramatic personal change, one she credited to a religious experience. In her remorse over what she had done she became a born-again Christian. Some called it a "jail house" conversion, not uncommon in persons condemned to die. Whatever the truth, the outward result was impressive. Gone was the cocky, hot-headed, draw-a-line-in-the-sand-and-I'll-show-you person of her youth. In its place was a gentler woman, convincingly remorseful and ferociously intent on surviving and helping others for whatever life she had left. She began with what would become an extensive correspondence with other women prisoners, offering emotional and spiritual support. In time she met Dana Brown, her prison minister, fell in love, and in 1995 married him. Although she had no hope of ever leaving prison, her life was full, and she wanted to live.

Few would deny how well Karla Faye represented her cause. Before she was finished, the Parliament of the European Union, the United Nations, and even the Pope had requested her life be spared. Her supporters included media figures such as Larry King and Charles Grodin. She had a compelling interview on *60 Minutes,* and conservative televangelist, Pat Robertson—a staunch advocate of the death penalty—publicly railed against her future execution on his widely disseminated *700 Club* program.

It didn't hurt that Tucker had become quite an attractive woman with a much softer appearance than years earlier during her trial. Dressed in crisp prison-whites, her brown hair pulled back in a ponytail, her dark eyes sparkling, she appeared younger than her actual age. With a warm and engaging smile, when she looked straight into the camera and said she was sorry, she came across as thoughtful and sincere. It was a terrible thing she had done, she admitted. She would spend the rest of her life—whatever remained—trying to make peace with her own conscience. Nothing she could do or say could bring back the two people. She often referred to them as "my victims," but she insisted the murderer she had been was now a changed person. She had found Jesus, and he had saved her. That's why she wanted to keep living, even if it meant residing in prison for the rest of her life. She thought she could do some good. Now 38 years old and on death row, she requested in as many ways as she could that her life be spared.

On Monday, February 2, 1998, the Texas Board of Pardons and Parole voted unanimously, 16-0, *against* clemency for Karla Faye Tucker. The result was no great surprise. The Board's previous year's tally for clemency in 16 death penalty cases was 266-0 against. As for the Governor, George W. Bush, Jr., a likely Republican nominee for President in 2000, he made it abundantly clear where he stood. Gender would have no place in his decision. Karla Faye Tucker had received her day in court. There was no question of her guilt. Now she must pay with her life. Despite his avowed support for special prison rehabilitation efforts, Governor Bush found Tucker's religious conversion and good works irrelevant to the question of clemency. As he saw it, there were only two legitimate questions: Did she commit the crime, and had she been afforded all due process she had coming to her? Nothing else mattered. The Board Chairman echoed the Governor's sentiments. Karla Faye's changed life carried no weight.

As a way of convincing the U.S. Supreme Court that appropriate capital jury guidance was being provided, Texas lawmakers had mandated *high probability* of *future dangerousness* as an absolute requirement for imposing the death penalty. Retribution alone was not sufficient reason for putting a criminal to death. (Most behavioral experts are quick to admit limited success in predicting future behavior, violent or otherwise. For twelve jurors to make such a prediction about any criminal is more an act of faith than a reasoned prediction regardless of the testimony of paid behavioral consultants.)

At the time of Karla Faye Tucker's trial, the jury's impression of her as a threat to society, while hardly scientific, was understandable. Could a person who committed a pickax murder with sadistic sexual overtones be trusted not to do something violent again? Not surprisingly, the jury's answer was no. But Tucker wasn't executed in a few weeks or months or even a few years. Her appointment with the lethal injection team was delayed for well over a decade. In the meantime she had become a different person, and it appeared a sustained change. By the time her execution date rolled around, it seemed highly unlikely she would be a repeat murderer in prison or out. (Despite the passage of almost fifteen years, no official reevaluation of her threat to society was ever undertaken.)

Another Changed Life

Leopold and Loeb settled into prison. In time they worked through their twisted feelings of superiority and became seriously involved in educating other prisoners. With the Warden's blessing they opened a correspondence school for inmates, drawing on other educated prisoners as teachers. Eleven years later, in 1936, Loeb was killed in a shower room razor attack during which he sustained more than 50 slash wounds. Leopold was devastated, but he continued their work and expanded his own self-education. He became an X-ray technician in the prison hospital, mastered 27 languages, reorganized the prison library, volunteered for an experimental malaria vaccine, and designed a new prison education program. He learned Braille, solely for the purpose of aiding a prisoner who had been blinded in the act of committing a crime.

In his later book, *Life Plus Ninety-Nine Years*, Leopold described how remorse for what he had done was "a constant companion... never out of my mind." Darrow wrote and visited Leopold several times, the last occasion, a year after Loeb's death, when Darrow was age 80 and less than a year away from his own demise. Leopold described the man to whom he owed his life: "The mask of death was on his face. But age and illness had not dimmed that piercing inner light. His wisdom, his kindness, his understanding love of his fellow man shown out as brilliantly on this last day I saw him as it had on the first" (Baatz, 2008).

Count Down

It was noon, Tuesday, February 3, 1998, when the final word came. The United States Supreme Court had rejected Karla Faye

Tucker's final request for a stay of execution. By mid-afternoon the countdown was well underway at Huntsville, Texas, her execution scheduled for 6:00 p.m.

"The Walls," as it's called, named after the towering red brick structures separating prisoners from the outside world, is located one block off the main drag in this small East Texas college town. Along the main street, shop window advertisements reflected the happening: "Karla Faye Tucker Sale—Killer Prices, Deals to Die For." By the early afternoon a sizable crowd of five or six hundred persons had gathered outside along with a mass of broadcast trucks from various outlets all over the world. Up front, closest to the prison, college students dominated the scene. Periodically, without any warning, they broke into loud chants. "She sliced, she diced, and now she's got to pay the price," or lengthy runs of "Kill her... kill her... kill her..." The raucous chants drowned out scattered calls for mercy. An assortment of hand-painted signs, mainly pro-execution, jutted up from the crowd like flowers in a field. *"One Last Orgasm, Juice Her." "Ax And You Shall Receive." "Forget Injection, Use A Pickax." "Glad You Found God, But Satan Found You First,"* and *"Have A Nice Day, Karla Faye."* Despite the chants and deadly sentiments, the student contingent seemed more intent on having a good time—a break from studies, an afternoon of fun in the sun and death jokes. What could be better?

Though less vocal, the anti-death penalty people carried their signs as well: *"Execution is No Solution,"* and *"Jesus loves Karla Faye, and So Do I."* Others, seemingly of no particular persuasion, wandered through the crowd, seemingly content simply to be near the action.

Easily recognizable from his numerous television appearances, Karla Faye's appeals lawyer David Botsford, dressed in a dark suit and tie, stood out in the crowd. Mustached with hair lines starting to recede, he peered intently through stylish glasses, working the media, giving interviews to anyone who asked. His message was upbeat. Several last-minute appeals in the works. Governor Bush had given his word, nothing was going to happen until the Supreme Court handed down its *final*, final ruling.

All along Botsford's main argument had been with the clemency process itself. He characterized it as purely cosmetic. The Board of Pardons and Parole never interviewed petitioners face to face. They hardly ever met all together as a group and never in public. And other than cases of proven innocence or evidence of gross injustice, they never recommended clemency. Since 1993 there had been 76 requests for clemency. *All* rejected. Clemency was supposed to be about mercy and compassion, he insisted, but not in Texas. He was not the first lawyer to challenge the state's approach to clemency, but so far the state had been unrelenting.

The Rest of the Story

At what proved to be Leopold's last parole hearing, the poet, Carl Sandburg, addressed the Parole Board. "He was in darkness when he came here," Sandburg said, "but he has made a magnificent struggle toward the light." It was left to Leopold to make his final plea. "Gentlemen, it is not easy to live with murder on your conscience. The fact that you did not do the actual killing does not help. My punishment has not been light… But the worst punishment comes from inside me. It is the torment of my own conscience. I can say that will be true the rest of my days… All I want in this life is a chance… to find redemption for myself by service to others. It is for that chance I humbly beg" (Baatz, 2008). After 30 years in prison, numerous parole hearings, and unrelenting, high-pitched emotional public debate, in 1958, Nathan Leopold walked out of prison a free man.

Attempting to escape the publicity generated by the newly released film, *Compulsion* (based on the Bobby Franks murder),

Leopold with the Parole Board's blessing traveled to Puerto Rico, bought a dog, and settled into a sedate life, out of the public eye.

After working for a year as a hospital technician (room and board, plus ten dollars a week) at the Church of Brethren Hospital, he finished a master's degree. He taught mathematics in Spanish, participated in research at the University of Puerto Rico Medical School on leprosy and intestinal diseases, and wrote an ornithology book — *Checklist of Birds of Puerto Rico and the Virgin Islands*. In 1961 he married a social worker, Trudi Feldman, and by all appearances lived a satisfying life. In 1963 he was granted formal release from parole which allowed him to drive an automobile, stay out late at night, and most important for him, travel the world outside Puerto Rico. Eight years later, after being hospitalized for a chronic heart condition, his wife at his side, Loeb died of a heart attack at age 55.

Ten minutes of seven. The sun had already set. Still, there was no word until off to the side a television crew pointed toward the front of the prison where several persons hurried down the front steps of the main building. "That's it," one of the men shouted, "there go the observers. It's all over." A few minutes later, the media announced to the world: "Tonight in Huntsville, Texas, eight minutes after receiving a lethal injection, Karla Faye Tucker was pronounced dead at 6:45 p.m."

In her final statement, Karla Faye apologized to the relatives of Jerry Lynn Dean and Deborah Thornton and thanked prison officials for treating her well. Her last words were reserved for friends and her husband: "I love all of you very much," she said. "I will see you all when you get there. I will wait for you." Reportedly, when the injection started, she closed her eyes, and half a minute later, after sighing softly, fell silent (Verhovek, 1998; Ward, 1998).

Richard Thornton, the husband of the murdered woman, witnessed the execution. "This was not fun and yet it was necessary," he said. "I want to say to all victims in the world: demand this. Don't ask for it. This is your right." But the murdered woman's brother, Ronald Carlson, who at Karla Faye's request also witnessed her execution, conveyed a different message. After several years of exchanging letters and visiting Karla Faye in prison, he had come to know her well. He called her execution "a pure atrocity."

On the radio, a woman accountant from Boston—a supporter of capital punishment—spoke in a broken voice. Karla Faye's execution troubled her deeply. She had followed the case closely. "It would have been easier if she had died right away," she lamented, "instead of waiting 13 years and having her change into a whole different person. I think that's what made it harder."

After Thoughts

Was Karla Faye Tucker more deserving of ultimate punishment than Nathan Leopold? Both committed horrific murders. The difference was not so much in the crimes as it was in the inconsistencies of adversarial justice. Darrow viewed capital punishment as society's way of finding a scapegoat and ignoring the real causes of capital crime. He insisted criminals, even the most vile, were also victims. Despite its outward appearance, the murder of Bobbie Franks, he said, was not a case of willful evil. Darrow was a powerful and persuasive communicator. One-on-one with a single judge he succeeded in making a compelling case. There were powerful and invisible forces over which the two boys had no control. The judge was persuaded, and Leopold and Loeb escaped execution.

Karla Faye Tucker was not so lucky. She had no Clarence Darrow. While well meaning, her trial lawyers were not in the same league. They failed to emphasize possible mitigating effects of a severely deprived life and heavy drug use while the prosecutor portrayed these experiences as evidence of abject evil and pictured Karla Faye as a continuing threat to society. Had the state been required to reevaluate Tucker for "dangerousness to society," likely she never would have been executed.

Defenses against blame are as old as blame itself. In the next chapter we review some of them starting with one of the oldest—*insanity*. By clarifying the minimal requirements for *criminal culpability*, the insanity defense opened the door to other blame defenses, more recently called mitigating factors. As it turns out, they are only precursors to the neurobehavioral case against blame itself.

Chapter 4

The Devil Made Me Do It and Other Defenses

We're depraved on accounta we're deprived.
Stephen Sondheim

Given our strong inner sense of self-determination, feelings of guilt often arise when we fail at what we are trying to do or when we break rules. When we observe the same in others, we project guilt onto them in the form of blame. Given that punishment has arisen as a response to blame, there is a proclivity for us to try and justify or explain away what we have done. Excuses abound some of which have been codified in the law. One of the oldest blame defenses is to claim that an unsound mind has suspended our free will and taken away our ability to control what we are doing.

Mad or Bad

In the spring of 1998, a disgruntled subordinate guard shot and killed the newly appointed Commander of the Pope's Swiss Guards and his wife before turning the gun on himself (*The Telegraph*, 2018). The official Vatican announcement described these killings as "a moment of madness." Perhaps so, but in this country, the question would remain: was this "mad" killer insane? Madness and insanity are not the same. Strictly speaking, "insanity" is a *legal term*—not medical—used officially in legal proceedings. This confusing relationship between madness and insanity can be summed up this way: while most insane persons are severely mentally ill, only a tiny percentage of mentally ill persons meet the legal criteria for insanity. This holds true even for persons with the most severe psychotic forms of mental illness such as schizophrenia. More often than not, delusional persons who in the throes of acute psychosis commit crimes do not qualify as insane. It's insanity, not madness, that provides an escape from blame. This puzzling discrepancy has a long and tortuous history.

The English/American history of insanity started with it being defined as a defect in judgment so severe a person could reason no better than a "wild beast." Under the law it took this degree of derangement for a person to escape blame. This crude common sense definition persisted until 1800 when an assassination attempt was made on King George III—who himself later would eventually become crazed (Robinson, 1996; Reznek, 1997). As the King entered the Royal Box of the Drury Lane Theatre in London, James Hadfield, a victim of two severe war wounds, tried to shoot him with a pistol. Although Hadfield's aim was widely off the mark, he was promptly arrested and charged with treasonous attempted murder.

At trial six weeks later, Hadfield was represented by Thomas Erskine, considered at the time England's greatest trial lawyer. Suspecting his client would fail the "wild beast test," Erskine made a creative leap when he proposed a less stringent definition of insanity; one requiring only that the person exhibit some form of *disordered thought*. This fit with evidence Hadfield was acting on the delusional belief that unless he sacrificed his own life, God would destroy the world. Believing suicide a mortal sin, he decided instead to kill the king as a substitute. While acknowledging his client understood perfectly well the wrongfulness of what he was doing, Erskine insisted madness need not make men complete idiots in order to excuse their actions. Disordered thinking was enough.

Given this considerably expanded definition of insanity, the jury's acceptance was surprising. A hastily arranged Act of Parliament provided a special verdict—*"not guilty by reason of insanity."* This allowed Hadfield to escape hanging in exchange for an indefinite commitment to the Bethlem Asylum. (In other words, not guilty but committed anyway more likely than not for the rest of his life.) This shell-game-aspect of the insanity defense—not guilty, but not really—persists to the present day, revealing the powerful resistance to accepting persons as blameless under any circumstances.

Forty years later the insanity defense underwent another dramatic change. Without any obvious reason, Daniel M'Naghten, a Scottish wood turner, abruptly abandoned his work and wandered off through Europe hearing "voices" and becoming increasingly paranoid. Eventually, having become fearful of being attacked by the Tories (at the time led by English Prime Minister, Sir Robert Peel), M'Naghten stalked Peel for days before mistakenly shooting and killing his secretary. When apprehended, M'Naghten explained his actions to the police as his way of fighting back against scurrilous accusations and threats on his life.

At M'Naghten's trial various so-called "mad doctors" offered up conflicting opinions as to what constitutes insanity (some things never change), but it was the lead medical witness, Dr. Edward Munro with his idea of *irresistible impulse,* whose testimony stood out. He insisted it was an overpowering loss of emotional control that caused M'Naghten's murderous assault without the failure of other mental faculties (Maeder, 1985).

Finding the doctor's assertion absurd, the presiding judge admonished the jury to ignore this fanciful idea of irresistible impulse and find M'Naghten guilty. Instead, after deliberating less than two minutes, the jury did the opposite, declaring the defendant "not guilty by reason of insanity." This decision, at least momentarily, greatly expanded the insanity defense's shield against blame.

Reining It In
The M'Naghten decision unleashed a huge public outcry. A

murderer was "getting off." (Not unlike what transpired much later in this country when John Hinckley was declared insane after shooting President Ronald Reagan.) In a quick response, the British House of Lords drafted a questionnaire on insanity and criminal responsibility. Bypassing the "mad doctors," the Lords submitted their questions to fifteen judges of the Queen's Bench, the answers to which would exert enormous influence on the future of insanity laws. The fact that these were judicial not medical opinions explains in part today's disconnect between legal insanity and severe clinical mental illness.

For starters, the judges declared the M'Naghten decision a grievous error! In a rollback of previously well-established opinion, they claimed the defendant was being given a free pass. He should have been found sane and hanged. While accepting that M'Naghten's delusion might well have influenced him, the judges insisted this was irrelevant. If the man understood what he was doing was *wrong and against the law*, nothing else mattered. Nuance counted for nothing. As for Dr. Munro "irresistible impulse" (monomania madness), the judges rejected the idea out of hand. Pure nonsense, they said. *Defective reasoning, alone,* was the proper basis for an insanity claim. Regardless of how strong the compulsion to commit a crime, if a person understood the difference between right and wrong, he was not insane and therefore fully blameworthy. With this ruling, insanity as a defense against blame was greatly narrowed in scope once again.

The M'Naghten Rule assumed individuals are free agents capable of choosing to live or not live within the law. As such when they commit wrongs, it's fully on them. To acknowledge irresistible impulses as insanity would be to open the door to blamelessness far too wide. How could any claim of *impulse-induced* crime ever be disproved? All crime would become acts of insanity; all prisons and jails would be replaced by asylums. There would be no basis for legitimate retribution. With their decision, the judges did their part in preventing blame excusing insanity from breaking out all over.

Still, though stricken from the law, the rudimentary idea of behavior being influenced by elements beyond a person's control survived and over a hundred years later resurfaced in the 1955 American Law Institute's *Model Penal Code*. This modern update of insanity declared a person not responsible [blameworthy] for criminal behavior if he lacked either the capacity to understand the criminality of his conduct *or* the ability to conform his behavior to the law. For a time this re-expanded concept of insanity was on the books and widely accepted, but there were complications in its implementation. In courtrooms, wildly conflicting "expert" opinions became standard fare. The task of deciphering after the fact a person's intention and state of mind at the time of the crime proved insurmountable and eventually led many observers to label these attempts pseudo-science.

This set the stage for yet another reversal. In 1982 when President Reagan's would-be assassin "got off" by pleading madness, all hell broke loose; despite convincing evidence that John Hinckley suffered from schizophrenia and acted out of a delusional belief, immediate public outrage erupted over his having "beat the wrap." A clamor to get rid of or severely rein in the insanity defense erupted once again. Congress swiftly replaced the Model Penal Code definition with a much narrower version (*Insanity Defense Reform Act*, 1984), which once again eliminated irresistible impulse and, in addition, shifted the burden of proving insanity to the defendant. Any hint of "avoidant" behaviors, such as trying to escape detection, disposing of evidence, or resisting arrest—even in cases where the defendant was blatantly psychotic—would constitute proof of the person's having understood the wrongfulness of what he or she had done. Disordered thinking from drugs or alcohol had no standing. Only when psychosis arose from a "mental disorder" could it be used as a defense against blame and only then if the person totally failed to appreciate the criminality of his or her act. Alone, crazed psychosis with command hallucinations, bizarre beliefs and even fears of threats to one's own life were not enough. This much narrower version of insanity as blame defense has held sway ever since.

Our current prevailing legal definition of insanity fails to recognize the parallel lives psychotic persons live: one preoccupied with delusional beliefs, hallucinations, and misguided moral convictions; the other engaged in normal day-to-day judgments and activities such as stopping at red lights, buying groceries, paying bills, fixing meals, cleaning up after kids, and showing up for appointments. This is what makes the psychotic experience so unpredictable: craziness living in tandem with rational behavior. What's right one moment is wrong the next, setting the stage for psychotic persons sometimes being *compelled* to do things which a part of them knows are "wrong."

Andrea Yates provides a poignant example (O'Malley, 2002). At age 37 this Texas woman, a previous honor student and champion swimmer, suffered a series of severe post-partum depressive episodes. On four occasions she was admitted to a psychiatric hospital. When at some point her doctor inexplicably stopped her antipsychotic medicine, she grew increasingly paranoid and as a result eventually ended up killing her five children. Later, she explained why. It was a deluded attempt to thwart Satan. Speaking to her out of the walls of her home, he told her he was about to abduct her children and take them to hell. Convinced this was true, she frantically worked out a plan which to her at the time seemed perfectly reasonable. On awakening one morning she fed her five children breakfast and with steely resolve proceeded to "save" them one by one by drowning them in the bathtub and then carefully laying them out on the bed *so they would reach heaven before Satan could abduct their souls*. Finished with these "mercy" killings, she called the police and calmly reported what she had done; how she had waited until no one else was in the home before methodically carrying out her gruesome task.

Appearing on Court TV's *Mugshots: A Mother's Madness—Andrea Yates*, Dr. Lucy Puryear, a Baylor College of Medicine psychiatrist, said of Yates: "She was the sickest person I had ever seen in my life" (Court TV, 2002). Even so, despite her plea of insanity, Yates was found sane and guilty of multiple murders.

(Later, as a result of flawed testimony by a forensic psychiatrist who incorrectly claimed Yates had copied her crime from an episode of television's *Law and Order*—an episode that never existed—her guilty verdict was set aside and she was declared "not guilty by reason of insanity." She was lucky. In most insanity defense cases the fact that a defendant acted on a delusional belief counts for nothing.)

Yates' official label of "not guilty" proved a mixed blessing. She was immediately admitted to a forensic hospital. After ten years of incarceration, when she asked to visit a church, a judge denied her request, and in 2014 her doctors—following a crush of critical comments from the media and the public—pulled back a petition for her to attend supervised events such as picnics. Although she hasn't seen the public for almost 20 years, Yates regularly creates artwork sold in the Kerrville State Hospital in Texas. Those persons who purchase her art are unaware of who created it. The money goes into a memorial fund used to promote mental health screening of low-income women.

Shell Game

In Hinckley's case what the public perceived as his "getting away with it" turned out to be something else. Instead of being allowed to go free, Hinckley fell into the Catch-22 world of not-guilty-but-not-really. His freedom was made contingent on his being judged no longer dangerous. And how did that work out for him? Hinckley ended up spending 35 years confined in a psychiatric hospital in Washington D.C. even though for most of this time his doctors insisted he was no longer dangerous. In 2003, even after a federal judge confirmed Hinckley was no threat, he was only allowed to spend 17 days a month at his mother's home, always under close surveillance. For an additional 13 *years* he remained an involuntary hospital patient until finally in July of 2016 U.S. Federal Judge Paul Friedman ordered his release.

On September 16 John Hinckley walked out of St. Elizabeth's Hospital a "free" man. Even then he was restricted to living with his aging mother, remaining within 50 miles of Williamsburg, and reporting regularly to a mental health team for treatment. In addition he was barred from talking to certain individuals and to the media generally.

In 2014 the National Association of Mental Health Program Directors reported the length of prison stays for persons found "not guilty by reason of insanity" averages 5-7 years; this despite evidence that persons found not guilty by reason of insanity have a subsequent arrest rate roughly *half* the general population. It's easy to see why some have characterized the promise of "not guilty by reason of insanity" as little more than a cynical legal ploy aimed at making sure his so-called defense against blame remains an empty promise.

In truth mental illness is but one of a complex host of factors contributing to criminal behavior, some genetic, some experiential, all strictly determined. Some have to do with how people are treated as they grow up—the opportunities or lack thereof they have, the degree of support, their role models, their exposure to poverty and abuse. But regardless of the causes, serious crimes are serious crimes. They require persons be held *responsible* and subject to appropriate consequences but to automatic imprisonment for beyond what is necessary.

In 1983 the American Medical Association voted to abolish the insanity defense altogether. To date, four states have done so. And why not? When you think about it, should a person who kills out of delusional conviction be given more of a pass than a strung-out, petty thief who murders someone in a robbery gone bad? Both individuals have broken the law, committed serious crimes, and until proven otherwise must be considered future risks.

Currently, the insanity defense's main legal role is to serve as a misleading legal symbol of how—short of the most extreme instances of brain failure—people act as free agents, fully blameworthy for crimes they *choose* to commit. The continuing assumption of illusionary blame with carved out defenses such as "insanity" serves only to perpetuate blame and punishment as appropriate responses to rule breakers as opposed to holding them *responsible without blame*.

The Quagmire of Degrees of Guilt

In earlier law, guilt was not quantified. Guilty was guilty. Regardless of the circumstances or a killer's state of mind, homicide was punishable by death. The insanity defense did succeed in cracking open the door to the idea of different *degrees of guilt*. Consider the act of taking someone's life. The charge will depend on a number of factors. Except for the relatively recent addition of "aggravated" murder, the current classification of homicide closely follows one outlined by William Bradford, Attorney General of Pennsylvania, in the late 18th century (Bradford, 1968).

Key to Bradford's classification is *intent*. Degree of guilt depends mainly on the amount of *deliberation and willfulness* involved in the crime. Under Bradford's system, the most blameworthy form of homicide is *premeditated, willful, and malicious* killing. This is *first-degree* murder. In cases where there is malice *without* premeditation, blame decreases. For example, if a person kills in a fit of rage, this is considered *second-degree* murder. If malice cannot be proven, blame is scaled down further to *voluntary manslaughter*.

An even lesser degree of blame is assigned when a person kills not willfully but as a matter of *negligence*. No deliberation, premeditation, or malice is required; only extreme recklessness. This is *involuntary manslaughter*. In some states, there is yet another category with even less severe consequences, variously called *reckless* or *vehicular* manslaughter. In this highly mobile age, apparently we consider killing someone with our automobile less blameworthy than killing by other means.

The degree of criminal blame can also be influenced by intoxication, but only partially. Intoxication's standing in the court has always been shaky. Early Anglo-American law categorically rejected drunkenness as a criminal defense. The reasoning was straightforward. Anyone who became intoxicated did so voluntarily. The court saw a direct connection between the "involuntary" loss of control from drunkenness to the voluntary choice to drink. Full blameworthiness was in order. (Some jurists even argued such crimes should be considered "aggravated" since they reflected *moral depravity*.)

Eventually, this position came under attack. The question was asked: what about the person who becomes so intoxicated he lacks the capacity for criminal intent. Regardless of his choosing to become inebriated, *without a guilty mind* (intoxicated) at the time of the crime, there is no basis for conviction. This line of reasoning created a legal dilemma. On the one hand, the acceptance of intoxication as a *complete* defense might encourage drunken crimes; on the other, faced with the growing acceptance of alcoholism as a medical disease, the idea of drunkenness as strictly a matter of choice became less tenable.

What to do? As a way out of this dilemma, jurists struck a compromise over the nature of intent. They arbitrarily declared that certain crimes required more focused attention than others. This led to the differentiation of two kinds of criminal intent: *specific* and *general*. If a certain crime was thought to require specific intent, and the charged person could prove that at the time of the crime he was incapable of forming such, he could not be convicted.

With respect to intoxication, jurists insisted while intoxication had no effect on general intent, it could compromise specific intent and by doing so provide a *partial defense* for certain crimes. While it would not exonerate unlawful breaking and entering someone's home, it might absolve blame for stealing items from the premises since this crime required specific intent. (If you are experiencing this legal argument as over-the-top convoluted, not too different from counting angels on the head of a pin, you would be correct. Such are the machinations of assigning blame and trying to defend against it.)

In truth this whole scheme is a legal fiction. There is no clear demarcation between specific and general intent. Even so, this rickety legal compromise is employed as a way of shading various degrees of blame.

The same can be said for *mens rea* (a guilty mind), a pillar of criminal law totally apart from considerations of insanity. Criminal convictions of any kind, in addition to requiring proof of an actual criminal act (*actus rea*), require the defendant have a guilty mind as well; one capable of understanding the nature of what is being done. As a result it's not uncommon in criminal proceedings for both sides to marshal "experts" for the purpose of deciphering the defendant's precise state of mind, not when the expert interviews him but at the time of the crime. To liken this to crystal ball gazing would be to give it far more credence than it deserves, filled as it is with speculative inferences, absent any objective facts. Yet this legal ritual persists. The side with the most persuasive "expert" wins, even in matters of life and death, despite the irrationality of the whole procedure.

Blameworthiness is a core legal assumption. Because people are viewed as free agents, they are judged blameworthy for crimes they commit. Only with great reluctance are exceptions made. Only when a person's ability to willfully and intentionally choose is perceived as compromised, does legal blame decrease and then only very reluctantly. This may occur as the result of insanity, intoxication, temporary emotional outbursts, or dementia. If it's determined a person has committed a crime totally bereft of free will (rarely), blame is eliminated altogether but as we have seen with the caveat of requiring sometimes years of close scrutiny and restrictions.

Our laws view free choice in the form of willful intent as a requisite to criminal blame. In light of our growing understanding of the vast number of invisible influences on human behavior, this position becomes increasingly problematic. What will the law and the courts do with solid evidence that this basis for assigning blame is illusionary?

Excuses vs. Explanations

As knowledge of genetic and experiential influences on behavior grows, the role of self-determination recedes. Ironically, this has opened the door to a myriad of legal excuses in the form of pop psychology explanations. In the hands of some practitioners, these fanciful explanatory justifications have proven convincing to jurors. But there is no consistency. What works in one trial, fails in another.

In August 1989, only a few days before they would murder their wealthy parents, Erik and Lyle Menendez, ages 22 and 19, drove from Los Angeles to San Diego (Davis, 1994). Using fake I.D.'s they purchased two guns and then returned to Los Angeles where they slept over at their parents' Beverly Hills mansion. The next day they went on an all-day, shark-fishing trip together, but only two nights later the brothers burst into their parents' television room and blasted them with their recently purchased 12-gauge shotguns. For the death-dealing, last shots, they held the guns directly to their parents' heads and fired. Later the defense would characterize the excessive shooting as "overkill," evidence of the panic the brothers allegedly felt. This was difficult to square with their having purchased movie tickets to establish an alibi before calling the police to report their parents' death, allegedly murdered by intruders. With $700,000 of life insurance money and the promise of a 14-million dollar estate, Erik and Lyle took off on a wild shopping spree, but suspicions quickly surfaced as to their possible involvement in the killings. Finally, when the therapist whom they had told about the crime reported being threatened, the brothers were taken into custody and charged with murder. After persisting in their denials for three years, a week before the first trial, they confessed.

Their main defense was a claim of sexual and emotional parental abuse. Both recounted stories involving sexual assaults on the younger brother by the father, including forced oral sex and sodomy. The defense characterized the killings as a product of "emotional sickness" arising out of years of abuse and constant fear. In the first trial, almost half the jurors accepted the defense's version of the crime: the unfortunate result of justified anger and fear, but a split over the question of murder versus manslaughter left the juries deadlocked (there was a different jury for each brother), and a mistrial was declared.

During a second trial, the abuse defense did not play nearly as well. Jurors saw it as a "fictional" excuse designed to justify a terrible crime and escape punishment. Erik and Lyle Menendez were convicted of first-degree murder and sentenced to life in prison.

The use of simplistic, pop-psychology explanations of crime has become routine courtroom fare. Speculative excuses masquerade as true defenses against blame. One psychiatrist, writing for *The Wall Street Journal*, described an exchange with a poorly educated burglar seeking help for the ill effects of an abusive childhood. When the psychiatrist inquired as to *what* specific ill effects, the man scratched his head and said: "Well, doctor, I keep on burgling." When the doctor suggested a more plausible reason, namely, to get money to buy things he wanted, the man angrily rejected such a "ridiculous idea." The doctor ended the article with a comment on how quickly pop science hits the streets and gets picked up as excuse material.

One close observer of criminal defenses likens the recent flood of courtroom excuses to a pseudo-disease which he calls "brain overclaim syndrome" (Morse, 2006). Harvard law professor Alan Dershowitz devoted an entire book to the subject of legal excuses. He called it *The Abuse Excuse: And Other Cop-Outs, Sob Stories, and Evasions of Responsibility* (Dershowitz, 1994). Statistical links between criminal behavior and specific abuses do not constitute proof in any individual instance. Take, for example, the potential influence of birth complications and parental rejection on future behavior. It is true these two adversities frequently turn up in the life histories of criminals. In a study of more than 4,000 eighteen-year-olds, Professor Adrian Raine, a University of Southern California psychologist, found children with both adverse experiences racked up *three times* the amount of violent crime as other teens (Raine, 2013). But detractors were quick to point out the obvious: most abused children who suffer birth complications grow up to be *nonviolent*. Using early life experiences as an automatic alibi for criminal behavior is questionable. Think about it. An enterprising lawyer with a serial killer for a client builds a defense by referencing an in-depth FBI study of serial killers that shows a majority were abused as children. The lawyer claims that, as with other serial killers, his client's unfortunate history of abuse made him kill. As farfetched as this might seem, versions of this defense are employed commonly in courts across the country, sometimes successfully.

In his book, *Explaining Hitler: The Search for the Origins of His Evil*, Ron Rosenbaum relates a myriad of "excusing" explanations applied to Hitler (Rosenbaum, 2014). They include his abusive father, an incestuous attachment to his mother, "post-encephalitic sociopathy," syphilitic brain disease, a genital injury, various forms of mental illness, and deep-seated anti-Semitism in the Germanic culture. Rosenbaum refers to an episode from the television series, *Unsolved Mysteries* (November 1991). Portraits of three "diabolical minds"—Ted Bundy, John Wayne Gacy, and Adolph Hitler—are presented. The bottom line explanation for why Hitler did what he did came down to this: "He subjugated and killed millions because he could not overcome feelings of inferiority." In other words, Hitler's behavior in Nazi Germany was due to *low self-esteem*.

Back in the 60s, the psychologist, Walter Mischel, evaluated four-year-olds enrolled in preschool on the Stanford University campus. Most of the subjects were children of Stanford faculty, staff, and graduate students. The study looked at how these children handled a certain situation (Mischel, 2014). During an interview, each child was told something had come up unexpectedly, but if the child waited for the 20 minutes it would take the researcher to run a quick errand, he could have two tasty marshmallows as soon as the researcher returned. If he couldn't wait, he didn't lose out altogether. He got *one* marshmallow, and he could eat it immediately. The researchers viewed this somewhat fiendish "marshmallow test" as a measure of *impulse control*.

Based on the outcome, possibly they were correct. As high schoolers those kids who had been able to hold out for two marshmallows proved better able to handle frustrations. Under stress, they were less likely to become rattled; and, in personal disputes, they were calmer and more patient. The "I-can't-wait" group—about a third of the original sample—had a far less attractive profile. *As a group*, they were more easily upset and prone to arguments and fights. While the study is fascinating, the average findings say nothing about any *particular individual*.

Why? Because many of the "impulsive" children *did not* experience problems later. Being at risk statistically does not automatically result in an adverse outcome.

Much of what behavioral experts present in courts is based on this kind of statistical illogic. More often than not, behavioral defenses are speculative excuses masquerading as scientific explanations. They offer one-size-fits-all justifications that could be applied equally as well in hundreds of other cases. Why such defenses win the day in certain cases but fail in others is a mystery, but one thing is certain, it is not due to the validity of the arguments.

Not too long ago mothers were still being blamed for their child's schizophrenia; PMS was considered a "hysterical" elaboration; and sexual impotency (pre-*Viagra*) was viewed as a sign of deep-seated psychological conflict. In her book, *Illness as Metaphor*, writer and philosopher, Susan Sontag, poignantly described her distress at being blamed by others for her own breast cancer (Sontag, 1978). With the emphasis on mind/body connections swirling around her, she found herself fending off accusations masquerading as laments: If only she had eaten better and managed her stress differently, she would not have contracted cancer. Too bad. In the absence of definitive causal understanding, blame thrives and excuses abound.

When Blame Is Set Aside

In the morass of largely unexplained human behavior, however, there are important exceptions. There are instances of criminal behavior thoroughly understood. These cases have important implications for the future of blame. They demonstrate how a deeper understanding of the causes of a criminal act override the assumption of free agency so that blame is set aside. Consider the following three, real-life examples of violence (Taylor, 2007).

A powerfully built man abruptly attempted to kill his wife and daughter with a butcher knife. After this failed attempt, the man appeared enraged. He snarled and kicked when anyone approached. His behavior was totally out of character, his wife insisted. She had never seen him this way before. After further questioning by a physician turned up a recent onset of severe headaches and blurred vision, subsequent neurological studies revealed a tumor in the anterior-temporal lobe of the man's brain. After the tumor was surgically removed, there were no further episodes of violence.

Having watched him remove his clothes and expose himself for the second time in a week, a frightened woman reported her neighbor to the police. The man was taken into custody as a sex pervert. The case was strengthened by a police report from four months earlier describing him standing motionless by the side of the road, naked from the waist down. When his wife was questioned, she recalled another incident when without any explanation her husband removed his clothes in front of their children. "He looked confused," she said. A more detailed medical history revealed that at age 17 the man had been knocked unconscious by a falling tree. Based on this information, an electroencephalogram was done which showed a characteristic complex-partial seizure pattern. Once started on an anti-seizure medication, the man's strange bouts of "exhibitionism" disappeared.

Without warning a middle-aged, academic administrator trashed the furniture in his home and threatened his family. This behavior was so out of character he was sent to a mental hospital for evaluation, but no diagnosis was made. Over the next several years, the man gradually experienced notable intellectual decline and became increasingly paranoid. The initial episode of violence was replayed numerous times. Eventually, he started beating his wife. Finally, after his physical health began to deteriorate, he was admitted to a general hospital where doctors diagnosed thyroid failure. Following treatment with replacement thyroid hormone, the man staged a dramatic recovery and returned to his previous work, fully recovered.

Until the medical causes of their violence became known, these three men were considered fully blameworthy and appropriate for punishment. Serious criminal behavior is rarely the exclusive result of brain tumors, seizures, or hypothyroidism. More commonly, the causes are far more varied, complex and beyond our comprehension, a mix of genetic and experiential influences. The causative map of human behavior resembles a giant puzzle with most of the pieces still missing. Still, as we shall see the broader picture of how genes and experience interact to determine behavior is relentlessly falling into place.

Having explored an array of blame defenses in this chapter, I hasten to add they are beside the point if what we will explore in Chapter 7, *The Astonishing Illusion*, is true. If what the neurobehavioral sciences are telling us about the narrative self and the after-the-fact nature of human experience is true, the basis for blame for anything and the justification for punishment disappears altogether. Proffered defenses are irrelevant. We are getting there.

But before we do, we will take a look at what our blame-and-punishment based system of criminal justice looks like currently. It's not pretty. Apart from the challenge we take up in Chapter 7, there are already substantial reasons for changing the way we approach rule breakers in this country which we will consider in the next chapter.

Chapter 5

The Scourge of Massive Incarceration

It isn't true that convicts live like animals:
animals have more room to move around.
Mario Vargas Llosa

It's difficult to come up with much of anything positive to say about prisons. The most intrusive elements of our criminal justice system, they are inhumane, costly, ineffective (at reducing recidivism), and generally void of any rehabilitative benefits. They emanate from the deeply-seated belief that rule breakers freely choose to do what they do and consequently deserve to be punished as a matter of correcting moral imbalance and as a deterrence against future rule breaking. Later, we'll see how both assumptions are false.

Massive Hell Holes

Through excessive use of force, regimentation, and ritualistic disrespect prisons operate as dangerous, demoralizing hellholes. The basic idea seems to be stuff as many rule breakers as possible inside prison walls and make it as inhumane as you can and still get away with it. The more overcrowding, the better as a part of rendering what these rule breakers deserve.

In 2011 a U.S. Supreme Court decision (*Brown v. Plata*) ruled mass incarceration in California prisons a violation of prisoners' eighth amendment right *prohibiting cruel and unusual punishment*. But as is often the case nothing much changed. Prisons remain our country's heavy-duty gold standard approach for achieving criminal justice (Deitch, 2018; Kelly, 2018).

Each year 12 million persons pass through our jails and prisons. For many of them (despite being nonviolent and low risk for absconding) imprisonment begins even *prior* to their being found guilty of a crime simply because they cannot make bail (Bazelon, 2020). Of major countries in the world, the United States has the *highest* rate of incarceration: *716 persons per 100,000 population*. Compare our rate to Iran (287/100,000) and Russia (439/100,000). And even more revealing is how we stack up against Western European countries: the UK (146/100,000), France (103/100,000), Germany (78/100,000), and Norway (74/100,000). Over the past 40 years the amount of U.S. incarceration has *quadrupled*, historically unprecedented and internationally unique (National Research Council, 2014). In 2012 close to 25% of the world's total prison population resided in U.S. prisons; this despite a population of only 5%.

There are obvious reasons for this state of hyper-incarceration in our country. Over the past several decades laws have been passed designed to increase the required amount of prison time for lesser offenders. In the 1980s a spate of them mandated lengthier prison sentences. This was followed in the 90s by half the states passing versions of the "three-strikes-you're-out" law mandating at least 25 years in prison for those who qualified with little regard for the seriousness of their offenses. In the same vein new laws required prisoners to do no less than 85% of their formal prison sentence. The result was the entrenchment of mass incarceration as the main tool for controlling crime despite its demonstrated ineffectiveness.

If you spend any time in one of our prisons—regardless of why—it's highly likely you will get to know the harsh, inhumane routine (Johnson, 2018). Without the enormous emphasis on submission, punishment, and deprivation, authorities would be forced to address real differences and needs in various rule breakers with individualized approaches. As it is, court monitors established through class-action suits are often required to insure even minimal therapeutic services. Small changes that could make huge differences are not made.

Inner Workings

The numbers only tell part of the disturbing story. At the core of massive incarceration is religious-like dedication to punishment for the sake of punishment. The amount of time a person is imprisoned is seen as only part of the penalty to be paid. Unofficially, deprivation and cruelty are thrown in for good measure. Out of sight, out of mind, the true character of these institutions flies under the public radar.

If you have never imagined yourself in prison, it's worth doing. They are despicable places filled with daily dangers, deprivation, disrespect, and absence of privacy. Dressed in their institutional clothes, inmates are herded like cattle from place to place according to rigid schedules. Mindless routines and the demand for absolute deference to authority—regardless of how off base—is beyond any questioning.

Solitary confinement where an inmate spends 23-hours alone in 9 x 5 cages for prison infractions is still used. Sometimes for years! You get to shower, and on select days, you get to move to a slightly bigger cage that resembles a dog run. Actual human contact is almost nonexistent. It's difficult to see how this practice escapes qualifying as torture (Benforado, 2015). And, sometimes, after suffering such extreme deprivation for many years, prisoners are released directly to the streets with little or no preparation. It happens.

Asleep or awake, safety is never assured, your life in danger every hour of the day. Physical assault is a constant threat from staff as well as other inmates. (Sexual assault from inmates and prison staff are roughly equal.) The threat may come from individual vendettas, psychotic rages, macho expressions of dominance, or gang action. You can never be sure when, where, or from what direction attacks on your body, mind, and life are coming. Unsurprisingly, suicide rates are elevated in prisons and jails.

Designed Centers of Violence

Based on scenes from an Alabama prison, Shaila Dewan opens a window into this pervasive air of prison violence in her article, "The Violence of Prison: A Rare, and Troubling Look Behind the Walls" (Dewan, 2019). "The contraband is scary enough: Homemade knives with grips whittled to fit in particular hands. Homemade machetes. And homemade armor; with hooks and magazines for padding. Then there's the blood. In puddles. In toilets. Scrawled on the wall in desperate messages. Bloody scalps, bloody footprints, blood streaming down a cheek like tears. And the dead: A man kneeling like a supplicant, hands bound behind his back with white fabric strips and black laces. Another hanging from a twisted sheet in the dark, virtually naked, illuminated by a flashlight beam."

Working as a consulting psychiatrist in a high security state prison, on one occasion I was scheduled to evaluate a new transfer. The brief report I had received noted how the man, in a fit of homophobic rage, had savagely knifed to death a gay cellmate. For this he had received a prison transfer for "reassessment" and a new placement. But within three hours of being celled with a new inmate—before I could see him—the prisoner was overheard in the chow hall boasting how he had just "snuffed" a queer. Guards were alerted and quickly confirmed his story after they found his cellie's dead body covered by a blanket. Whoever was in charge of designating new cell assignments had failed to take into account the killer's history and the fact his new cellmate was gay. Constant threat is a part of basic prison culture.

And there's no escaping the obvious racial element involved in mass incarceration. In her book, *The New Jim Crow: Mass Incarceration in the Age of Colorblindness*, Michelle Alexander provides this stark snapshot: "The United States imprisons a larger percentage of its black population than South Africa did at the height of apartheid. In Washington, D.C.... it is estimated three out of four young black men (and nearly all those in the poorest neighborhoods) can expect to serve time in prison" (Alexander, 2010).

When it comes to prison, if you are a person who made a mistake, wants to cause no trouble, do your time, and get out, good luck. Only those who are quick learners make it. Adaptation is key to prison survival, even if it means breaking rules, joining a gang, fighting, or taking advantage of others. I once had an inmate tell me: "If you aren't a hardened criminal when you come here, you damn well are by the time you leave."

One Size Fits All

While some decisions regarding prison housing are based on level of violence or gang affiliation, for the most part prison/jail life is a matter of everyone being thrown together: extremely violent with nonviolent, predators with mentally disordered persons, multiple repeat offenders with first timers. In 44 states more persons with mental illness reside in jail or prison than in the largest remaining state psychiatric hospital. There are three times the number of serious mentally ill in jails and prisons than in hospitals (Treatment Advocacy Center, 2016). With the closing of state institutions across the country in the 1960s and 70s we have gone from roughly 600,000 mentally ill persons in mental hospitals to 350,000 in prisons and jails with 250,000 homeless and on the streets (Frances, 2018).

The May 18-20 weekend edition of *The Wall Street Journal* carried an article denoting similar statistics and promoting a new generation of institutions designed especially for chronically mentally ill patients (Husock, 2018). These "new models" are inspired by non-institutionalized, therapeutic work-and-living communities of the late 1800s and early 1900s. Few familiar with the grossly inappropriate criminalization of the mentally ill can argue with the need for radical change. It's long overdue. The current situation is unfair both to patients and criminal justice staff.

I have worked as a psychiatrist in several prisons, including one with highest security (Pelican Bay in California). Many of the patients I saw expressed fears of harm or even death. Most were bunked in unbelievably small confines; typically, two men to a cell with two bunks, a toilet and sink. No privacy, the total square footage living space extremely small, sometimes less than 100 square feet. Inmates lived with only a single window (placed in the door), covered from the outside on the whim of any guard. When I saw prisoners in individual psychiatric sessions, regardless of their official diagnosis, it was clear most of them struggled with the prison culture itself. It was a constant stress, and not surprisingly, anxiety was a major complaint as was difficulty sleeping. Most were pill seeking, partly related to how psych pills (of any kind) were valuable currency in the prison's unique barter system. But the pills also gave relief to how they felt being captive in a dangerous setting with few places to hide.

As I walked prison halls and yards (scary exercise itself), the regimented nature of prisoner lives was constantly on display. For the most part inmates moved en masse in groups at appointed times. I never tasted the food, but it was a source of common complaint except for the fresh fruit which when available could be confiscated and made into "pruno" (a prison-made alcohol drink). In time inmates who showed themselves highly responsible escaped some of the harsh discipline by becoming building "tenders." Under the oversight of prison personnel they were allowed to perform work essential to the operation of the prison.

In one prison, as part of a mental health program I helped establish, I taught a beginning writing course. Well attended, it was only available to prisoners with mental health diagnoses. Several commented on how much it was a welcome respite from a life of almost jungle-like existence. I remember my own emotional "shutter" on one occasion when the realization hit me of what it would be like to be housed in prison and how probably I would not survive very long. Typically, inmates were presumed potentially violent, serious escape risks, and fully deserving of the punishment they got. Those with "mental illness" were often viewed as "freeloaders" by much of the staff and other inmates. The constant clanging of cell doors and opening and closing of sally ports, the predictable takedowns each day, the frequent shutting down of cell blocks because of "prisoner counts" in error or the discovery of contraband in a person's cell—all of these together made the prison a massive pressure cooker. And being imprisoned in a SHU took it to a whole different level.

SHU (pronounced shoe) stands for Special Housing Unit. It's the short-hand name for administrative units where inmates are placed in solitary confinement. This happens for a variety of reasons, from serious violence to lesser criminal acts such as stealing and repeat insubordination. Typically, this means 22-24 hours a day alone in a cell with a concrete slab and mattress. A couple of times a week the person is allowed a 15-20 minute shower. When exercise is permitted, it amounts to pacing up and down in a narrow concrete "box." In the state of California there had been no limit on how much time could be spent in a SHU until recently when a cap of 5 *years* was instituted. (The United Nations has deemed a period longer than 15 days torture.) In the highest security prison I worked, I was told roughly 1/3 of all prisoners were kept in solitary confinement. When I saw SHU inmates as patients, they would be shackled, handcuffed, and placed in a wire cage. In turn, I would be outfitted with a bullet-proof vest. I had no authority to question the practice, but clearly it was not a very therapeutic setup.

Private prisons—which now account for confinement of 8% of the country's inmates—have been promoted as a possible answer to the ravages of mass incarceration. Results suggest otherwise. Shane Bauer in his book, *American Prison: A Reporter's*

Undercover Journey into the Business of Punishment, reports his own first-hand account of life in a privately-run prison. In his four-month tenure as a prison worker he was aware of dozens of stabbings, scores of "use of force" incidents, and routinely sub-par medical care. Prison guards described for him how sometimes they used inmates as prey in training bloodhounds. He was also told about beatings conducted outside the view of security cameras. Bauer explains how management decisions to reduce expenses more often than not were dictated by a search for greater profitability. As a result pay for prison staff competed with Walmart (Blakeslee, 2018).

Recent reports detail how "overflow" undocumented immigrant detainees are now being housed in prisons. In court-filed documents, Lisa Hay, Oregon's Chief Federal Defender, described the abysmal conditions in a Sheridan, Oregon prison complex commandeered for "detainee" service. She reported detainees triple bunked in cells measuring 75 square feet, "subjected to strip searches," and confined to their cells for approximately 22 hours a day.

Small deprivations pile up in these punitive cultures. Consider the simple act of making phone calls as a way of keeping up with friends and loved ones and trying to arrange for life after prison. For years prison inmates in Texas prisons were charged exorbitant rates as they were forced to pony up 26 cents for every minute of a phone call. For the limited time allowed (20 minutes), the total cost came to $5.20, money not easy to come by for many of these prisoners. It was not until September 2018 that the Texas Board of Criminal Justice got around to addressing the problem. Under new guidelines inmates will now be allowed 30-minute phone calls and charged $1.80 instead of what under the old rate would have been $7.80 (*Dallas Morning News*, 2018).

With respect to voting rights, only two states (Vermont and Maine) allow imprisoned felons to vote. Denied voting rights are restored only in 20 states after prison, parole, and probation; in three states after prison and parole. At the conclusion of prison sentences alone they are given back in 15 states and the District of Columbia (ProCon, 2018).

There was a time when prisoners could join class-action suits to protest their living conditions and basic rights, but less so in recent years as a result of court decisions that make prisoner class-action suits much more difficult to file.

On the false promise of achieving great reductions in prison populations, criminal justice budget cuts and staff reductions have been made, but attempts to fill these positions with far less trained personnel have met with difficulty (Reilly, 2018). A *USA Today* report entitled "Nurses, Cooks Enlisted as Guards" describes untrained persons pulled into duty without meaningful orientation or even appropriate uniforms. "Hundreds of secretaries, teachers, counselors, cooks and medial staffers were tapped last year to fill guard posts across the Bureau of Prisons because of acute officer shortages and overtime limits, according to prison records..." (Johnson, 2018).

After Prison Purgatory

The scars of the mass incarceration experience carry over after prison. Most inmates leave with little money, no place to stay, no job, and nonexistent social support. Previous relationships are either busted or toxic. To make matters worse, by law released inmates have to "check the box" when they apply for jobs, indicating they are an "ex-con." And in most instances the mandatory parole experience only makes matters worse. Filled with mindless rules and "gotcha" regulations, this alleged supportive oversight becomes the gateway to prison return. Technical "violations" such as missed appointments, hanging out with the "wrong" people, "dirty" urines, and even bad attitudes become the ticket back.

In the political arena numbers of inmates become impressive

scoreboard tallies. But for the vast majority of imprisoned criminals, it is only a matter of time before they walk out of prison and hit the streets again having been demeaned and brutalized in the prison/jail experience and having learned little more than how to be better criminals, only now with added rage. Roughly three out of four will be back in three years.

Summary Judgment

Penal institutions should be straight-out denounced as inhumane and ineffective. As it is, their paramilitary, oppressive culture is tolerated mainly because the majority of prisoners are poor, minorities, and marginal individuals with little political clout. Current penal practices are sadistic and misguided, off-the-mark substitutes for badly-needed educational and therapeutic services. Like a throwback to earlier centuries, hardened criminals, the mentally ill, the addicted, the developmentally challenged, and the poor are all thrown together and treated the same. To cap it off, some states and the federal government continue to use death as punishment despite clear evidence of its being erroneously handed out, non-deterrent, and demonstrably biased. The most recent *Death Penalty Report* (Amnesty International) lists only nine countries that regularly kill persons as punishment, and we are one of them: U.S., China, Iran, Iraq, Saudi Arabia, Somalia, Yemen, Sudan, and North Korea (Amnesty International, 2017).

So, this is where the blame illusion has taken us. The assumption of free choice leads us to believe "intentional" crimes deserve punishment in punishing places. While the excessive focus on being tough and punishing as much as possible has been good for politics, it pushes aside time and effort that could be put to use reforming, educating, and making job ready the vast majority of prisoners who otherwise will at some point leave prison at high risk for returning.

Our current penal system is costly, unimaginative, and hung

up on punishment. *Without any reference to what we are about to explore in Chapters 6 and 7, there is a powerful case for a top-to-bottom redesign.* Despite longstanding criticism, change has been hard to come by. Recent passage of legislation in the U.S. Congress (First Step Act, 2018), as laudable as it was in reducing penalties for certain federal drug offenses, barely touches the surface. Nothing significant happens as long as mass incarceration paramilitary style continues to be equated with essential "law-and-order." The portrayal of prisons and jails as bulwarks against all that is bad sends a false but comforting message to a public reassured by the trademark practices of harsh discipline, primitive housing, inhumane deprivation, and token rehabilitation. In Chapter 10, Blameless Justice, we will consider a starkly different alternative.

But in the next chapter we shift gears to consider the *genetic and experiential (environmental) causes* of who we are and what we do. You'll note there is no inclusion of *free will* for a simple reason: despite our compelling personal convictions, there is no evidence that free will is anything other than a deep-seated illusion. What and who we are is the product of genes and experience. Full stop. Period. The implications for criminal justice are far reaching.

Chapter 6

Intricate Dance: Genes and Experience (Environment)

Once you see the forces that govern behavior, it's harder to blame the behavior.
Robert Wright

What causes us to do what we do? Is it our genes? The way we are brought up? Where we were born, the order of our birth, circumstances of our childhood, parental influence, peers, our education, traumatic events, socioeconomic factors? In a manner reminiscent of holy wars, the intellectual fighting over this question has been fierce in what has been called the *nature/ nurture debate*: genes vs. experience (environment) including our personal experiences as well as our surroundings—physical, cultural, social, and economic (Wright, 1997). From time to time spurts of new knowledge enable one side of the debate to surge ahead only to be overtaken later by newer evidence more supportive of the other side. Attempts to totally vanquish the opposition have failed repeatedly.

Today few purists exist on either side. The shrillness of the debate is gone. In his book, *Lifelines*, British biologist Steven Rose describes this new truce: "The phenomena of life are always and inexorably simultaneously about nature *and* nurture, and the phenomena of human existence and experience are always simultaneously biological *and* social. Adequate explanations must involve both" (Rose, 1997). This modern compromise synthesis represents a shift away from viewing genes and environment as either/or propositions and more toward understanding the two influences as inseparable—gene/ experience—as they work together in intricate combinations to

100

make us who we are.

Genes

The 17th century British philosopher, John Locke, penned a famous piece entitled "An Essay Concerning Human Understanding" (1689). In it he dismissed any innate human knowledge present at birth. For Locke it was clear, we enter the world a blank slate to be filled with experience, sensation, and reflection. He could not have been more wrong. In drawing this conclusion he was leaving out the extraordinary information-dense genes we inherit from our parents, fully in place as we enter the world.

While this chapter makes clear genes are not the whole story, their influence is dynamic and massive, much of it outside our awareness. In his book, *Genome: The Autobiography of a Species in 23 Chapters*, Matt Ridley sums it up this way: "Genes are not puppet masters or blueprints. Nor are they just carriers of heredity. They are active during life. They switch each other on and off; they respond to the environment" (Ridley, 2004).

Ridley drills down on the structure and workings of genes, characterizing the human genome as a book comprised of 23 chapters (chromosomes). Each chapter consists of thousands of genes arranged in numerous paragraphs (exons) but interrupted periodically by "advertisements" (introns). Each paragraph is composed of words ("codons") which in turn are made up of letters (bases). While our own English language books are written in words of variable lengths drawn from 26 letters, the human genome is written exclusively in three-letter words, each composed of four basic letters (amino acids): A (adenine), C (cytosine), G (guanine), and T (thymine). There are approximately one billion words in the book (human genome) which runs roughly the length of 800 Bibles end to end. Reading this genetic book at a rate of one word per second for eight hours a day would take a century to finish (Ridley, 1999). Amazingly, this extraordinary encyclopedic genome fits inside the nucleus of a cell much smaller than the head of a pin!

Think about it. Genes provide the basic design for our bodies and minds. They store the core human program that unfolds first in the uterus and then in our lives as developing infants and children and finally adults. They determine our general physical structure and appearance. Provide the platform for language, and stealthily exert major influence on who we become—our personalities and intelligence—as well as altering our risks for various diseases, up and down. Richard Dawkins, the well-known British evolutionary biologist and author of *The Selfish Gene*, was so taken with the gene's power he finally concluded the human body was little more than a "gene-survival machine" (Dawkins, 1978).

In writing his history of the gene, Siddhartha Mukherjee arrives at a similar conclusion: "Human beings are ultimately nothing but carriers—passageways for genes," he says. "They ride us into the ground like racehorses from generation to generation. Genes don't think about what constitutes good or evil. They don't care whether we are happy or unhappy. We're just means to an end for them. The only thing they think about is what's most efficient for them" (Mukherjee, 2016). Though a tad cynical this description does capture the powerful influence genes have on our lives.

We live in an unusual time when the ability to modify our own genes has become a reality. With CRISPR/Cas9 technology, gene alterations can be made with great precision (Doudna, 2017). In the past such gene change has been limited to the painfully slow process of selective breeding. Now, for good or bad, we have the power to modify genes, insert them where they haven't been, delete them or change them out. Recently, several cases of sickle cell anemia—involving a single misplaced letter—have been cured by harvesting, reconfiguring, and then reinserting a patient's own stem cells. With genetic cloning, we can even copy whole organisms. The ethical implications rightly hold back unrestrained gene modification efforts, but with little doubt what we have seen so far ultimately will pale in comparison with what is to come.

Power of Single Genes

The power of genes is dramatically illustrated in conditions that arise solely as the result of a single gene. Such is the case with Huntington's disease, the expression of a single, dominant, non-sex gene. When inherited, a normal parallel gene supplied by the other parent provides no protection. In time this horrific disease emerges regardless of what transpires in the person's life, 100% of the time. Huntington's truly is a *genetic* disorder. (Luckily, as it turns out, such purely genetic conditions are relatively rare.) When at least one of the parents carries the gene, each child has a 50% chance of developing the disease. Unfortunately, the major symptoms of Huntington's— progressively uncontrolled tics and bizarre body twisting combined eventually with cognitive decline, psychosis, and gradual muscle failure that compromises speech, walking, and eating—are not usually manifest until middle age after children have been born. Extraordinary and humbling to think all this mayhem is the product of a simple coding error where three letters of microscopic DNA are mistakenly repeated in the composition of a single gene.

Another example of the power of a single-gene aberration is a mystifying condition first described in an isolated village in the Dominican Republic and later in the eastern highlands of Papua New Guinea. Families from both areas reported instances of dramatic "gender transformation" occurring at puberty (Conis, 2006). The mystery was not solved until the 1970s when Dr. Julianne Imperato-McGinley of Cornell University traveled to a remote part of the Dominican Republic and confirmed these hard-to-believe reports.

The villagers described how at birth the sex of these children appeared female. Accordingly, the children were raised as little girls and developed normally until as they entered puberty, their bodies abruptly changed. Their voices deepened, they became decidedly more muscular, and most unsettling they grew penises. While these children were rare (about 2% of the children in the region), the story was always the same: little girls unsuspectingly evolving into men. Later, psychological evaluations showed these gender-flipping persons making good adjustment to becoming men and eventually fathers. In the Dominican Republic they are called *guevedoces* which roughly translated means "penis at age twelve." Similar persons in the Sambian villages of Papua New Guinea are referred to as *turnims*, "those expected to become men."

After documenting these stories Imperato-McGinley and her team eventually worked out what caused this startling transformation: a single gene mutation causing a deficiency of the enzyme *5-alpha reductase*. At birth and through childhood, the low level of this enzyme prevented the conversion of testosterone to its biologically active form, dihydrotestosterone (DHT), and by doing so blocked the normal expression of male external genitalia. But at puberty as the deficiency was overcome by a torrent of testosterone, classic male physical characteristics burst out all over.

A different part of the world provides yet another striking example of what a single gene can do. It involves men who without any obvious reason routinely become violent. In 1978 working at the University Hospital in Nijmegen, Netherlands, Dr. Han Brunner was approached by a woman seeking genetic counseling (Raine, 2013). She related how several of her male relatives were prone to explosive rage attacks and serious violence. Now that her own son was showing similar traits, she wondered if there might be a genetic problem.

Impressed with the woman's story, Brunner undertook a

study of what turned out to be a large Dutch family. Going back four generations he conducted extensive psychosocial interviews and took blood samples. As the woman had told him, he found several males who typically overreacted to frustrating or stressful situations with menacing expressions, excessive rage, and violence.

It took Brunner and his team close to 15 years to sort out what was behind these instances of extreme male aggression (Brunner, 1993).The problem—subsequently called the "Brunner Syndrome"—was caused by an aberrant X-chromosome gene that normally codes for monoamine oxidase A. In 14 male relatives the gene defect caused a complete absence of this enzyme which breaks down key brain amines, including serotonin, dopamine, and norepinephrine. As a result, their amine levels were consistently elevated. In his *Science* report Brunner concluded: "MAO-A deficiency is associated with a recognizable behavioral phenotype that included disturbed regulation of impulsive violence." He also found these same men had substantially lower I.Q.'s than normal (averaging roughly 85), as well as higher rates of ADHD, alcoholism, and drug abuse.

Rarely, even recessive gene mutations alone produce major problems. Consider an aberration in the gene involved in producing the protein known as POMC. This defect leads to an infant from birth experiencing ravenous hunger. In most cases the child—often red haired and pale skinned—becomes severely obese by the age of one. The problem is lifelong. But it arises *only* if both parents have this recessive genetic mutation; otherwise, the infant has no symptoms (Epstein, 2018).

Polygenetic

But as powerful as single genes can be, most of the genetic influence expressed in our lives is of a *polygenetic* variety: several genes working together in concert with our life experiences (including memories). In this ongoing intricate dance involving genes and experience, the contribution of each is difficult to tease out. An index labeled *heritability* has been developed in an attempt to capture the relative contributions of genes and experience. Theoretically, a score of 0% would indicate no genetic contribution; 100% total genetic determination. Research determinations of heritability of various human behaviors and experience are often quite surprising. Nicotine dependence, for example, shows 60% heritability. Age at menopause, 47%. Left-handedness, 26%. And height a notable 86%. But even so, height has experiential influences shown dramatically by the divergent trajectories of North and South Koreans following the end of the Korean War. South Korea went on to rapidly become a dynamic economy (11th largest in the world) with a universal healthcare system. In contrast North Korea's economy stagnated into poverty and even starvation as more and more of its income was diverted into military products and operations. Today the average South Korean is a full inch taller than his or her North Korean counterpart, this despite similar gene pools (Zimmer, 2018).

Most of us have no trouble accepting the idea of genes influencing our height, facial features, eye or hair color. But personality and behavioral traits such as boldness, risk taking, or empathy, with these we have more trouble accepting the role of genes. Probably because they are more closely tied to our sense of self-determination. We find it even harder to accept a role for genes in our political preferences, religious convictions, and choices of life partners.

Identical Twins as Window into Genes

One way researchers have tried to establish heritability more precisely is by studying *identical twins*. The reasoning is straightforward. Since identical twins have the same genetic DNA (unlike fraternal twins who are no more similar genetically than other siblings), any differences found between them are more likely the product of experience than genes. But the apparent elegance of this idea is frustrated by practical complications. For example, twins raised in the same family, in addition to having identical genes, share the same social environment, making it impossible to differentiate the effect of one from the other. As a way around this obstacle some researchers have focused on identical twins *separated at birth*. Under this circumstance, given identical genomes, any differences in the two individuals could be presumed to result from varied experience.

For over 20 years, the *Minnesota Twin Study* headed by Dr. Thomas Bouchard has studied a range of behaviors influenced by genes far more than previously known (Bouchard, 1990). With respect to personality traits, estimates of *heritability* (genetic influence) are roughly 0.50 (1.0 total influence); that is, 50% of the determination of these features is accounted for by genes. Not family, not friends, not unusual life experiences. The estimated heritability of intelligence is even higher at 0.70. Surprisingly, genes also seem to play a significant role in a variety of personal idiosyncrasies such as hand gestures, pet-naming, and nervous giggling.

Similar findings are reported by British geneticist Robert Plomin in his book, *Blueprint: How DNA Makes Us Who We Are*. Based on decades of twin studies he finds the influence of genes even stronger: "... the DNA differences inherited from our parents at the moment of conception," he surmises, "are the consistent, lifelong source of psychological individuality, the blueprint that makes us who we are... DNA isn't all that matters but it matters more than everything else put together in terms of stable psychological traits that make us who are" (Plomin, 2018A). While Plomin credits environment with considerable influence on what we do, he is quick to point how what looks like the work of environment sometimes is often genetics in disguise. "Parents respond to their children's genetically driven traits, and children see, modify and even create experiences correlated with their genetic propensities" (Plomin, 2018B).

One final aspect of genes. Each is enveloped by a collection of molecules known as the *epigenome*. While the intricate workings of this structure remain largely unknown, what we do know is that it can switch its associated gene on and off as a part of daily cycles. There is also good evidence that epigenomes can be affected by experience so as to be turned on or off for the duration. Far more debatable is the question of whether or not such changes can be passed on to the next generation. While there is little doubt this occurs in plants, it remains an open question with animals (Zimmer, 2018).

Environment (Experience)

Even so, although twin studies have unearthed these intriguing findings of gene influence over things not previously considered "genetic," they have done nothing to put aside the fundamental truth: our lives reflect a complex interplay between genes and environment. Consider obsessive compulsive disorder (OCD), a condition marked by recurrent intrusive thoughts and impulsive repetitive behaviors. A person whose identical twin has developed OCD has a 50-60% chance of having the same problem. It's a big difference, but does that make OCD a genetic disease? Close to an equal number of twins will *not* develop OCD, presumably due to differing personal experiences (Eapen, 2015).

The influence of environment on genes is readily apparent with respect to cognitive traits such as IQ. For kids raised in affluent circumstances, the genetic/intelligence correlation is roughly 50-70%; for those growing up in lower socioeconomic families, 10%. It's assumed affluence typically allows for fuller expression of these genetic factors while low socioeconomic circumstances inhibit them. Such interactions of genes and environment lead biologist and author, Robert Sapolsky, to conclude: "... it's not meaningful to ask what a gene does, just what it does in a particular environment" (Sapolsky, 2017).

Despite the strong argument that genes and experience add up to the sum total of what we are, for the most part, the intricacies of how they interact remain largely unknown. What we do know is this: genes may enhance, diminish, or completely block the effects of experience, and the reverse is true as well. Consider the life of Stephen Hawking, the acclaimed physicist, who died March 14, 2018.

Hawking's death came *53 years after* his physician told him when he was 23 that he would be dead in two years. He suffered from a rare form of early amyotrophic lateral sclerosis (ALS, Lou Gehrig's disease) traditionally considered a genetic disease. Along the way Hawking lost movement in his limbs as well as his ability to speak so that eventually he required assistance, speaking, eating and even breathing. Typically, death follows within a few years of an ALS diagnosis, but Hawking proved a remarkable exception. Just as his disease appeared on the verge of doing him in, the process slowed dramatically, allowing time for him to make major discoveries about the universe, particularly regarding those strange entities known as black holes. Before Hawking's life ended, he occupied the same professor chair once held by Sir Isaac Newton; and, even though eventually he could only communicate verbally with a speech synthesizer, his mind remained brilliant.

The genetic component of ALS remains a mystery. Although 15 different gene candidates have been identified only 10% of cases of ALS show clear-cut genetic inheritance. It's a reminder of the continuing mystery of how genes and environment mysteriously interact to make us who we are.

Like genes, experiential factors are sometimes the whole ball game. If a faulty constructed overpass collapses just as you are passing underneath, genes are not a likely cause. If you eat in a reputable restaurant and contract food poisoning, again, genes probably aren't a major factor. Similarly, on the positive side, if you purchase your first lottery ticket and win the jackpot, genes can't be given much credit. But for most aspects of our lives, the causes of what goes on in our lives and directions they take are a combination.

Despite our strong conviction otherwise, as far as science knows genes and experience together account for everything we are and do. Most of how this takes place never enters our consciousness as anything other than a narrative illusion. In summary, our lives are gene/experience. Despite the compelling subjective belief we hold, there is no scientific evidence for free will being a causative factor in our lives.

In contrast to the uniform, microscopic packets of DNA that comprise our genes, our experience, including the environment around us, is sprawling. It starts with our life in the uterus, an environment that typically is safe and secure but one that can become unhealthy by things happening in our mother's body that surrounds us and on which we depend. Our environment includes past experiences and their memories which repeatedly influence us when then they come to mind. Our parents or other persons who raise us. The people we meet, the things we do. The challenges we face. The traumas we go through. Our successes, our failures. The books we read and increasingly the social connections we make through the World Wide Web beyond our immediate presence. The movies we see, games we play, the Internet we occupy, the art we immerse ourselves in. Even our introspective silences and subconscious meanderings. Our dreams. The earth we live on. The substances we are exposed to.

Mid-afternoon on July 18, 1984, James Huberty, a middle-aged man, walked into a McDonald's next to the Post Office in San Ysidro, California, carrying a 9-millimeter semiautomatic Uzi. Without saying a word he discharged 257 rounds into a sea of customers, ages 7 months to 74 years. Huberty was shot dead by a SWAT team sniper from a perch atop the Post Office next door. His autopsy revealed an unexpected finding. Hair analysis showed extremely high levels of cadmium (lead as well). William Walsh, the chemical engineer conducting the test commented: "... highest cadmium level we have ever seen in a human being" (Raine, 2013). In retrospect, the main reason for James Huberty's out-of-character act of violence seems obvious. Before quitting, Huberty had worked at Union Metal for many years. During his exit interview he gave his short but get-your-attention reason for leaving: "The fumes are making me crazy," he complained. Of course not all mass killers are *cadmium toxic*. The point is we live in a physical world of earth, air, and water that influences us in ways we are not aware of all through our lives.

The Body As Environment

Seldom do we think of the physical structure we live in—our bodies—as an environmental factor. But it is. It's our most immediate surrounding influencing us continuously. Extra tall, extra short, either can have far-reaching implications, particularly for sport and work. Being 6'8" may open the door to basketball but block the path to becoming a ballerina. Similarly, the color of one's skin or shape of one's eyes. Core body weight can open and shut doors; impede disease or make it more likely.

Equally as real are anatomical effects on our interpersonal lives. Appearance is a potent social currency. More deep seated are the neuroanatomical structures channeling our feelings of love, sex, and attachment (Fisher, 2016). When studied closely, behaviors that seem totally spontaneous—flirting, sexual stares, shadowing body movements—turn out to be traits hardwired into human brains.

Microbiome

Only recently have we become aware of the importance of other life continually residing inside us. In his book *I Contain*

Multitudes, Ed Yong rejects the idea that we have a single genome. "This is a pleasant fiction," he insists. "In fact, we are legion, each and every one of us. Always a 'we' and never a 'me'" (Yong, 2016). He goes on to explain how the "microbiome"—a gigantic microbe colony we carry around in our gut—contains roughly 500 times more genes (in more than 1,000 separate microbial species) than our own human genome (Mosley, 2018). Yong describes ways these extra-human genes play critical roles digesting food, producing certain vitamins, breaking down toxins, and crowding out potentially dangerous microbes or actually killing them with antimicrobial chemicals. This gut microbial colony stimulates the growth of organs in our body by releasing certain chemicals.

There is growing evidence that microbiome microbes affect our cognitive function, emotional well-being, and immune responses (Dinan, 2015). Presumably this results from ongoing brain-gut communication via an enteric nervous system (ENS) which in his book, *The Mind-Gut Connection,* Emeran Mayer describes as a *second brain* "made up of 50-100 million nerve cells, as many as are contained in your spinal cord" (Mayer, 2016). This auxiliary nervous system is awash in the same neurotransmitters found in the brain; in fact, 95% of the body's serotonin resides in the gut. The idea of acting on "gut feelings" may not be so fanciful after all based on gut reactions relayed to the brain via the vagus nerve and preserved as memories to be elicited by future challenges and stressors.

With respect to the risk of diabetes, the type of food a person ingests has been assumed the most important determinant of severe spikes in blood glucose levels, but a study of 800 non-diabetic persons suggests otherwise. Extensive testing of more than a hundred factors showed that food was *not* the critical determinant. That title went to the composition of the gut bacteria, the microbiome (Topol, 2019).

A recent announcement from *Synlogic,* a synthetic biology

company, provides a preview of future work related to the microbiome (*Doctor News*, 2018). By programming probiotic bacteria to carry out functions persons with certain metabolic disorders have difficulty managing, the company is creating "smart medicines." One product targets ammonia which in severe liver disease can reach toxic levels and cause severe brain dysfunction. Taken daily in pill form, the engineered probiotic moves through the gut metabolizing excess ammonia into a harmless substance. Similarly, other engineered probiotics will enhance the corrective role of the microbiome specific to each disorder (Bone, 2018).

Amazing to think how the entire working of the microbiome—like most of what is happening throughout our bodies—goes on outside our awareness or control. It reminds us of how much like an iceberg human consciousness and awareness are. Our strong conviction of being in control of our own lives fails to take into account so many things that affect us moment by moment. Would our conviction be otherwise if we knew the whole story?

Experiential Stressors

Life stresses can cause psychological problems but also major physical changes. A condition known as "psychosocial dwarfism" sometimes occurs in younger children as a consequence of *extended stress* in the home. Shortened stature is the direct result of the resulting suppression of growth hormone production by the pituitary. The effect lasts as long as the child remains in the stressful situation.

Social scientists have long recognized the major influence of socioeconomic status (SES) on human behavior as is well illustrated by a recent Harvard study (Chetty, 2016).

Research from a project called *Experiment Moving to Opportunity* looked at the impact of housing vouchers on the future of young children. For children under the age of 13, being able to move to a better neighborhood proved highly beneficial. As adults their incomes were 31% higher than were those children whose families failed to receive housing help. Where we live, how we live, at what time in history—these "uncontrollable" factors are potent out-of-sight-out-of-mind experiential influences on our lives.

Situation and Timing

The interaction of genes and experience is going on all the time outside our awareness. Some genetic programs require the presence of certain environmental factors *at a specific time* in order to turn on. Potent as they are, genes do not always get their way. Several years ago the psychology researcher Sandra Scarr-Salapatek studied the effects of genetics and environment on I.Q. (Scarr-Salapatek, 1971). She focused on 130 black and mixed-race-children adopted by *affluent* families. Compared to children raised in poverty, these children had I.Q.'s 10-20 points higher. The earlier the adoption, the greater the effect. Based on these differences, Professor Scarr-Salapatek concluded that the genes involved worked their influence best in more enriched environments. In their absence genetic expression remained either completely unexpressed or severely restrained.

The mysterious musical ability known as perfect pitch is another example of how exposure to a particular experience or environment can be critical to gene expression. A person with this remarkable trait hears a tone and knows precisely what note it is—C sharp, for example. Although mediated through genes, the acquisition of perfect pitch depends on being exposed to music *at a critical time*. Absent this, expression of the involved gene (or genes) is compromised. One study of 600 musicians who took music lessons before the age of four showed 40% developing perfect pitch in contrast to only 3% of persons whose lessons came after the age of 12 (Baharloo, 1998). Without the timely requisite experience gene expression proved incomplete or in many cases totally blocked.

In his book, *The Relationship Code*, George Washington University psychologist David Reiss reviews findings of a 12-year study of 720 pairs of adolescents with various degrees of genetic relatedness (Reiss, 2003). He concludes that many "genetic" factors are *triggered or suppressed* by a child's relationships with the important people in his or her life. Genes do not express themselves in a vacuum. Experience provides either a fertile or barren field or something in between. No experience escapes genetic influence, and gene expression is often modified by experience.

The metabolic disease PKU (phenylketonuria) dramatically illustrates gene/environment interaction. A person with this condition has a single gene defect that blocks a critical enzyme from birth. As a result, his or her body fails to metabolize the amino acid, *phenylalanine,* ordinarily a straightforward chemical reaction. A buildup results in a toxic chemical in the person's blood resulting in progressive intellectual impairment, seizures, severe behavioral problems, and early death.

If PKU is construed strictly as a "genetic" disease, there is no solution, but when viewed as a gene/environment problem, the answer is straightforward. Despite the defective gene, if the newborn's parents keep the child on a low phenylalanine diet, no buildup occurs and the problems of this otherwise tragic condition are avoided. Recently, two drugs have been approved to aid the breakdown of phenylalanine, making dietary restrictions far less demanding (Cunningham, 2012; Brooks, 2018).

So is PKU genetic or environmental? Clearly, it's both as is everything else about us.

Memes

The term "meme" was introduced by Richard Dawkins in his best seller, *The Selfish Gene.* The basic idea is this: similar to the way genes provide units of information for biological life, memes in the form of ideas, beliefs, cultural practices, and styles drive the social/psychological fabric of our lives (Blackmore, 1999).

Susan Blackmore refined and expanded the concept. "Memes spread themselves around indiscriminately without regard to whether these are useful, neutral, or positively harmful to us. A brilliant new scientific idea, or a technological intervention, may spread because of its usefulness. A song like *Jingle Bells* may spread because it sounds OK, though it is not seriously useful and can definitely get on your nerves."

Memes arise out of a fierce competition for our attention. For reasons not fully understood some prevail, others fall by the wayside. Some are amazingly quick in their spread and tenacious in their grip on our minds. Others, less so. More commonly we call them "fads." An essential aspect of the most influential memes is their degree of catchiness or stickiness. If you want a more personal introduction to memes, go through your old photos and check out your hair and clothing styles. Notice how they change over the years in concert with the people around you. Such is the stealth nature of *memetic* influence.

Memes spread like infectious agents. Some come and hang on for generations. Others are much more short lived. School shootings, fashion vogues, the narrowing of automobile styles, food fashions, "health" practices. With the emergence of social media, the speed and pervasiveness of memes is greatly enhanced. But as Blackmore suggests, memes can be productive, but they can also be destructive as evidenced by scams, pyramid schemes, cults, fake news, political campaigns, and health-compromising fads.

Unlike genes, memes are *Lamarckian* in the way they spread. The word derives from the French 19th century naturalist, Jean-Baptiste Lamarck. His view of evolution (later proven wrong) conflicted with Darwin's idea of natural selection. Lamarck was convinced physical changes *acquired during one's lifetime* would pass naturally to offspring. From a Lamarckian perspective, if a bodybuilder spent hours doing weights enough to develop prize-winning muscles, his oversize musculature would automatically pass to his children. Although with a growing knowledge of genetic transmission, this idea fell by the wayside, it's a perfect description of the way memes spread as they are transmitted non-genetically to others as a form of social influence.

We are discussing memes because they are powerful environmental influences on human behavior. In a *New York Times* article, "Catching a Ride On the Juul Wave," Amos Barshad describes a spreading practice known as *vaping*. First designed by two former Stanford students to mimic the action of a real cigarette, E-cigarette devices rapidly became the rage. Recently, Nielsen reported *Juul* sales had achieved a 54% market share of smokers, exceeding Marlboro cigarettes at the peak of their popularity. One senior high school class president commented: "What resonates with our generation are the memes... you'll see a bunch of memes about Juuling. It's just like making it more socially acceptable—it's perpetuating the thing that vaping is cool" (Barshad, 2018). Despite not producing tobacco smoke, evidence is rapidly accumulating for vaping having a gateway role to traditional smoking. Many Juul users were not previous cigarette smokers. While the nicotine content of a Juul started off in the 1-2% range, presently 4-5% is not unusual. Such concentrations approach the nicotine equivalent of an entire pack of cigarettes. More recently an alarming number of cases of serious pulmonary complications have surfaced raising serious health questions about this rapidly spreading meme.

A more prominent marketing meme is the SUV. Sales of this automobile design have skyrocketed forcing some major car manufacturers to discontinue more long-standing popular sedan lines to make room for more SUVs. This shift has occurred despite little evidence of any greater need for bigger vehicles.

Pursuit of SUV as meme seems a more likely explanation. Potential negative consequences are already in evidence. A *Detroit Free Press/USA Today* investigation recently implicated the "SUV revolution" in the escalating *pedestrian death rate*, up 46% since 2009, accounting for the deaths of roughly 6,000 pedestrians in 2016 alone, the approximate number of Americans killed in Iraq and Afghanistan combat since 2002 (Lawrence, 2018). With tremendous advances in meme-spreading devices, one has to believe the influence of this social version of DNA will only grow stronger. This is particularly true of the explosion of social media and its alarming capacity for "fake news."

Algorithms

A rapidly growing experiential influence on our lives are algorithms operating in the gigantic digital platforms of companies such as Facebook, Google and Amazon. In a much-watched TED talk Zeynep Tufekci, a technosociologist and professor of digital media, asserts that the tsunami of persons clicking on ads becomes the basis for what she terms *persuasive architecture*. By sifting through massive amounts of personal data bought and sold around the world individual vulnerabilities and susceptibilities are being identified. Highly accurate "probabilistic" guesses are being made concerning risks of addiction, religious views, degrees of happiness, and even sexual orientation. She illustrates with an example where "manic" persons were targeted for the sale of Las Vegas casino tickets. Tufekci claims even the persons overseeing these algorithmic programs do not understand how they accomplish what they do since strategies are constantly evolving as new data is analyzed and the AI program adapts. These algorithms are able to devise highly individualized messaging capable of influencing political decisions as well as making sales. A growing subtlety of this form of persuasive messaging seems inevitable (Tufekci, 2017).

Luck

In this chapter we have only touched the surface of experience/ environmental influence on our lives, much of it totally outside our consciousness. Among other things, we have not considered religion, culture, society, poverty, parental guidance or lack thereof, peer influence, addictions, unemployment, and a host of traumatic experiences—all powerful in their influence potential. But before we finish I want to mention one other factor: *luck*.

The acclaimed actor, Paul Newman, shied away from interviews, but when he did them he often insisted a key part of his success was *luck*, and you didn't get the idea he was being falsely modest. He was on to something applicable to life in general but perhaps not in the way he meant. When portrayed as random acts of fate without cause, lucky happenings are misrepresented. In fact winning the lottery, "chance" encounters, life-changing events, meeting "the right one," sitting down at a cafe next to a well-known celebrity, missing a plane that crashes, being born into a fabulously rich family, inheriting long life genes—none of these happen without cause. They only appear to do so because of the special significance they have for us, their unplanned nature, and our ignorance of their underlying causes. Daniel Dennett uses a much larger canvas to make the point: "There is a sense in which we are all... extraordinarily lucky to be here in the winner's circle. Not a single one of *our* ancestors suffered the misfortune of dying childless! When one thinks of the millions of generations of predecessors knocked out in the early round of the natural selection tournament, it must seem that the odds against *our* existing are astronomical" (Dennett, 1984).

Like all other happenings in our lives, lucky *and* unlucky events are strictly determined by factors mostly outside our awareness. In the material world of brain functioning, cause and effect is the unswerving, fully determined order of the day. Contrast this with our day-to-day subjective world conviction of being free agents, routinely overriding the influence of genes and experience as we free will and choose our way through life. How do we reconcile these two incredibly disparate views? We finally take up this question directly in the next chapter, *The Astonishing Illusion*. This is the point at which the reader needs to suspend disbelief, at least temporarily, and let the evidence speak for itself. Neurobehavioral science's version of the way things really are is truly astonishing.

Chapter 7

The Astonishing Illusion

By the time we think we know something—it is part of our conscious experience—the brain has already done its work.
Michael Gazzaniga

It's a universal conviction. Experiencing ourselves as free-willing agents, routinely overriding other influences as we contemplate, decide and act our way through life. To suggest otherwise prompts riotous disbelief. Impossible. Every minute of our waking day confirms what we "know." Preposterous for anyone to think otherwise. Isaiah Berlin, the Russian/British philosopher and historian, went so far as to assert that proof of the nonexistence of free will, if it should ever materialize, would be calamitous. Without self-determination, he insisted, traditional morality would be turned on its head, the basis for blame, eliminated. Social chaos would reign (Berlin, 1998).

We are convinced to the core of a self-determined nature unlike other forms of life or anything else around us. Asked how certain we are of it, most would give it a ten—as solid as waking up in the morning, deciding to make coffee or tea, and then doing it (or asking someone else do it!). Closely tied to this core conviction is an equally strong belief in the moral correctness of blame. Although, as a matter of defending ourselves in the moment we may try to downplay blame or escape it all together, deep down we know it goes hand in hand with freedom to choose. Life is what we make of it. As captains of our fate, it's on us, both the praise and the blame that go with it.

Surprisingly, despite this universal conviction, a great debate about it has raged for centuries. Samuel Johnson put it in perspective: "All theory is against freedom of the will;

all experience is for it." On one side of the debate is the inner conviction of what we know to be true from experience. Of course there is free will, the argument goes. Look, watch me pick up this book. Turn the pages. Mark my place. Put it back on the shelf. All of this, I choose to do. There it is: free will. Case closed. On the other side is the logical absurdity of such an idea: some ethereal, pixie-dust-using, genie decision maker hiding out somewhere in our brain, free willing and choosing with wild abandon in open defiance of the laws of cause and effect. More recently Harvard psychologist, Daniel Wegner, a man who before his untimely death from ALS at age 65 had scrutinized free will as much as anyone, boiled the question down to this: "Do we consciously cause what we do, or do our actions happen to us" (Wegner, 2002)?

The truth of the matter has a lot riding on it, but up until modern time there has been no way to resolve this centuries-long standoff between what we "know" emotionally and the illogic of free will.

Intimations of Trouble in Self Land

Predicting the future has always been a dicey proposition, mostly wrong. Still, over the years, science fiction writers have come up with some amazingly right-on-target guesses. In 1865 the French writer Jules Verne in his novel *From the Earth to the Moon* described manned space flight. Granted his idea of how it would happen (a huge spaceship-shooting gun) was off target; even so, a century later a spacecraft with the same name—*Apollo*—landed on the moon with the same number of persons aboard as appeared in Verne's novel.

Thirty-one years ahead of time, H.G. Wells predicted "atomic bombs" (*The World Set Free*), and in an 1888 utopian novel—*Looking Backward*—Edward Bellamy described inhabitants of a mysterious society using *credit cards*. During a 1964 BBC appearance Arthur C. Clarke, inventor and science fiction author of novels such as *2001: A Space Odyssey* and *Childhood's End*, described a future of instant communication through the use of space satellites as well as the rise of telemedicine. With such prescient predictions in mind, it's intriguing to find recent films, novels, and television programs questioning our deep-seated belief in free will and self-agency.

In Peter Weir's 1998 movie, *The Truman Show*, the main character at age 29 has an "aha" moment when he suddenly realizes his life is totally contrived. Sea Haven, his hometown, turns out to be a complete sham: a gigantic TV studio made to mimic the real world. His wife, family and friends, as it turns out, are actually hired actors from central casting reading scripted lines in a 24-hour reality show under the direction of an all-controlling TV director.

From his hidden control booth high in the ceiling of an artificial dome that doubles as the sky above Sea Haven, director Christof guides a group of set designers, grips, and actors as they portray Truman's every move for a worldwide television audience, 24-hours a day. Through it all, Truman has had no idea. The masquerade might well have gone on the rest of his life had it not been for a piece of studio equipment falling out of the "sky" on a clear day and landing at his feet.

This bizarre moment triggers Truman's awakening to his life as the orchestrated workings of someone else. Everything he has believed previously has been an *illusion*. Astonished and deflated, Truman finally decides to make a run for it, but Christof—well acquainted with Truman's water phobia— quickly counters by conjuring up a raging sea as an escape barrier. In the end Truman finds another escape route and finally walks away from his TV studio world into the unknown. (Readers of philosophy will note the similarity to Plato's Cave where the inhabitants believe the shadows on the wall are real, unaware of life outside the cave.)

In their 1999 movie, *The Matrix*, the Wachowskis portray

an even darker vision of an illusionary reality. Out of a troubling feeling that "something is not right," Neo, a low-tier computer hacker, finally tracks down Morpheus, a renowned cybercriminal and leader of a rebel group engaged in a struggle with advanced machines, the designers of the *Matrix*.

"Do you want to know what it [the Matrix] is?" Morpheus asks. "It is the world pulled over your eyes to blind you from the truth." He then recounts how all humans are now enslaved floating pods overseen by the machines who periodically harvest them for energy. This idea resembles the philosophical thought experiment known as "the brain in a vat" where a person's brain is removed from his or her body and floated in a vat filled with nutrient-enriched fluid. Nerve endings of the brain are connected to a supercomputer which with the aid of an extraordinary piece of software creates the illusion of human experience as it was before. People, objects, sights and sounds. Love and dejection. Successes and failures, all seemingly real as real (Putnam, 1981).

Morpheus wastes no time. He takes two pills from his pocket, one red, one blue. "This is your last chance. There is no turning back," he cautions before explaining how if Neo takes the blue pill everything remains the same: an illusion the machines have created. But if he takes the red pill, "You stay in Wonderland and I show you how deep the rabbit hole goes." Neo takes the red one and the deadly *Matrix* game is on.

A more recent portrayal of human experience *sans* free will appears in the television series, *Westworld*. Based on Michael Crichton's novel and 1973 film by the same name, the story takes place in a futuristic Old West theme park filled with android hosts (robots with human features) acting out various characters (Cookman, 2017). Customers who frequent the park pay hefty amounts to enter various storylines that allow them to interact with these expendable android characters. Virtually anything goes. The heart of the *Westworld* story follows the androids as they gradually rise up to free themselves from what they have come to recognize as programmed narratives. Having been designed to erase each encounter, two of the main android characters (a hardened town madam and a confused and distressed young woman) experience glitches that make them recall traumas they endure on a recurring basis. These recollections start them on the road to sentience. Will the androids of *Westworld* arrive at full consciousness?

In these various narrative worlds "the natives" grow restless as they come to suspect the possibility of life as it's experienced being *illusionary*. In this awakening they anticipate what neurobehavioral science is now revealing about ourselves.

Game Changer

If it exists, where does free will come from? How does it work? Is it a miniature person inside our heads pulling all the gears and levers like the Wizard of Oz behind the curtain? Unlikely. Apart from fantasy and fiction, nothing in the world as we know it happens without cause. A less than full understanding of how the brain works would seem a questionable basis for suspending belief in cause and effect and adopting in its place the idea of a free spirit directing our lives, yet this is precisely what we do.

Think about it. If true, how would an autonomous, free-willing self avoid a life of unpredictable twists and turns as it careens randomly from one willful intention to another? It's difficult to see how such random, free-willed, out-of-the blue behavior would lead to anything other than total chaos. But as implausible as it seems, this is the central belief on which our lives are based. More specifically, it is the bedrock, foundational assumption of traditional views of justice and morality. Persons go to prison—some even put to death—because they are thought to have chosen by their own free will to commit criminal acts.

Enter the neurobehavioral sciences with startling hard

evidence that free will is an *illusion*. Yuval Noah Harari, author of two masterful volumes, *Sapiens: A Brief History of Humankind* and *Homo Deus: A Brief History of Tomorrow*, characterizes these findings as "a time bomb in the laboratory" (Harari, 2011, 2015). "Today, when scholars ask why a man drew a knife and stabbed someone to death, answering 'Because he chose to' doesn't cut the mustard. Instead, geneticists and brain scientists provide a much more detailed answer: 'He did it due to such-and-such electrochemical processes in the brain, which were shaped by a particular genetic make-up, which reflect ancient evolutionary pressures coupled with chance mutations.'"

Without qualification, Harari characterizes free will as an illusion. He challenges the person who resists this shocking idea to carry out a brief experiment: "... decide not to think about anything at all for the next sixty seconds... just try and see what happens. Should be no problem," he insists, "if you indeed are master of your thoughts and decisions." I would add to Harari's test an additional simple observation. As you engage in conversation be aware of how most of the time *you* are not coming up with the individual words. They are flowing automatically *without any direction from you*. True, sometimes the flow stops or stutters and you consciously get back into the game looking for the right word, but then everything is back on track, and you are talking once again as though a ventriloquist is putting words in your mouth. Speaking turns out to be a window into what's really happening all the time, and it's *not* self-direction.

The Deceptive Brain

In *How the Mind Works*, evolutionary psychologist Steven Pinker describes our dual existence: "I ask, 'A penny for your thoughts?' You reply by telling me the content of your daydreams, your plans for the day, your aches and itches, and the colors, shapes and sounds in front of you. But you cannot tell me about the enzymes secreted by your stomach, the current settings of your heart and breathing rate, the computations in your brain that recover 3-D shapes from the 2-D retinas, the rules of syntax that order the words as you speak, or the sequence of muscle contractions that allow you to pick up a glass" (Pinker, 1997). The fact is the vast workings of our body and brain go on outside our awareness.

The human brain is one of the most complicated mechanisms known. This 3-pound mass of firm tofu-like material is continually processing billions of signals. With 100 billion neurons each connecting with 1,000 other neurons, it's estimated there are trillions of synaptic connections with thousands of signals being exchanged each second. Each signal is modulated by hundreds of different proteins and mediated by dozens of different neurotransmitters (Frances, 2013). All of this is part of a material reality with which we have no direct knowledge.

At the level of consciousness, even a tiny fraction of brain information would be a confusing overload. It would be like massive clutter in a garage filled with much more stuff than it can hold. Consider the way vision arises and you get a sense of how important it is to pare down the amount of conscious information. Light particles striking the back walls of our eyes (retinas) are translated into millions of electrochemical impulses. These impulses are then carried over networks of neurons to different nodes and relay stations where various syntheses and interpretations occur before arriving in the visual cortex at the back of the brain ready for a final translation. To shield us from this monstrously large amount of continuous streaming data requires it be converted into a much simpler and manageable subjective reality, say a rainbow or the smell of baking bread as opposed to electrochemical brain language. When we're conscious we are using the narrowest of bandwidths while all around us massive superhighways of information continuously flow bumper to bumper in our brains.

This idea of the world we live in being a stand-in for something else has been well explored by cognitive scientist Donald Hoffman. He puts it this way: "… what we perceive is never the world directly, but rather our brain's best guess at what the world is like, a kind of internal simulation of an external reality." He emphasizes the importance of this subjective perspective (user interface) as a fitness guide, not a conveyor of truth. "Evolution has shaped us with perceptions that allow us to survive. They guide adaptive behaviors. But part of that involves hiding from us the stuff we don't need to know. And that's pretty much all of reality, whatever [material] reality might be. If you had to spend all that time figuring it out, the tiger would eat you" (Gefter, 2016).

Thomas Campbell, a physicist, has another name for it: *a reality simulation*. Both men are in agreement: the human experience is something entirely different from what it seems. It is a simplified translation of a far more complicated reality (Hoffman, 2008; Campbell, 2017). Among other things, it gives rise to an astonishing illusion.

The Astonishing Illusion

In his fine book, *The Body*, Bill Bryson describes the visual *blind spot* we have as a result of bundled nerve fibers exiting the retina. This conduit—about the thickness of a pencil—creates what should produce a hole in our vision but does not. Why? Because the brain fills in the gap. "That's quite remarkable," Bryson concludes, "that a significant part of everything you 'see' is actually imagined" (Bryson, 2019). What's even more remarkable is the conclusion that neurobehavioral science drives us to regarding our entire human experience; *everything we experience* is in fact a translation of something entirely different into narrative form.

Quite amazing. Massive amounts of brain data reduced to a metaphorical narrative with each of us featured as central protagonist in a life story. So amazing that it goes largely ignored, particularly as it applies to matters of right and wrong. Neuroscientist and author David Eagleman describes the disparity this way: "There is a tension between biology and the law... After all, we are driven to be who we are by vast and complex biological networks. We do not come to the table as blank slates, free to take in the world and come to open-ended decisions. In fact, it is not clear how much the conscious *you* as opposed to the genetic and neural *you*—gets to do any deciding at all" (Eagleman, 2011). Still, courts and the law persist in clinging to an archaic blame-based psychology; one that assumes human behavior the product of self-determined choices. Best to keep the judicial doors shut to what science has to say than to rock the legal foundations.

Granted, when first considered this emerging view of the illusionary self seems outrageous, like something out of *The Twilight Zone*, but you be the judge. To the extent the evidence we are about to review is true, there's little doubt: our lives as we experience them are stories; free choice and the blame it inspires, illusionary actions of a storied character—that would be you and me. Paraphrasing Spinoza, the neuroscientist philosopher Sam Harris summarizes this narrative existence this way: "You can do what you decide to do—but you cannot decide what you will decide to do" (Harris, 2012).

How It Happens

The most compelling direct evidence for the "astonishing illusion" comes from research that tracks brain activity parallel with choices we think we make. As it turns out, what we sense to be "choices" are not really choices at all, but *only reports of what has already been decided* in the brain.

This is the central finding of work from the imaginative neurophysiologist, Benjamin Libet (Libet, 2000). In one of his most famous studies, subjects were seated facing a large clock (with a second hand) on the wall. After being wired with scalp electrodes for monitoring electrical brain activity, their straightforward task was explained. They were instructed to periodically (at their own choosing) decide to move a finger. The instruction was simple: "At random flex your forefinger noting the time of your decision on the clock-face monitor."

No surprise when results showed a predictable delay of approximately one-fifth second between the person's decision to flex his or her finger and the actual flexing itself. A person decides and then, after the brief time it takes to translate thought into action through nerve pathways, acts on it. Nothing new here. Nothing surprising. But Libet's second, unexpected and seemingly spooky finding, was a whole different matter. Roughly a third of a second *prior* to the subject's deciding to move a finger, the scalp electrodes consistently picked up a flurry of brain activity. This so-called *readiness potential* invariably preceded what subjects perceived as their own decision to act. Libet summed it up this way: "The brain 'decides' to initiate or at least to prepare to initiate the act before there is any reportable subjective awareness such a decision has taken place." In other words, the conviction of making a decision (choosing) was actually an *after-the-fact* perception conjured up by the brain as a fictional account for what had already happened!

In his book *Who's In Charge?* Michael Gazzaniga, a neuroscientist whose work on split-brain patients we are about to review, suggests a simple way of testing this seemingly outrageous idea of delayed experience. If you put your finger to your nose, you feel the sensations of nose and finger at the same time, a physical impossibility since neurons carrying nasal sensation are only about three inches from where it's processed in the brain compared to three and a half feet for the finger. Since the two impulses travel at the same speed, there is a difference in when the two sensations arrive, but we experience them happening simultaneously. Gazzaniga explains with respect to

Libet's work: "Consciousness takes time, but it arrives after the work is done!" (Gazzaniga, 2011).

Although Libet's name has become synonymous with the idea of *delayed awareness*, an earlier researcher made similar observations. Back in the 60s Grey Walter was hard at work as an EEG researcher. In his research on *choice*, participants were wired with electrodes to measure *readiness potentials* from the part of the brain controlling hand movements. They were then instructed on how to push a projector button in order to advance slides one after the other (Walter, 1964). What participants were not told was how these electrodes had been directly wired to the slide projector so that the brain readiness potential itself triggered the slides. As the experiment proceeded, subjects were amazed at how the slides advanced themselves just *before* they pressed the button. Another demonstration of the perception of *choice* occurring after the decision had already been made.

The Finnish psychologist, Risto Naatanen, pursued Libet's work with an intriguing variation. He had participants attempt to *fool* the brain (1978). As a way of putting thoughts of the research protocol out of mind, subjects were instructed to distract themselves by reading a book. The idea was to *trick* the brain into believing they were out of their decision-making mode. Then, according to plan, suddenly and without warning, they were to abruptly press a signal key in an attempt to catch the brain unawares. Ingenious as it was, the plan to "sneak up" on the brain failed miserably. Regardless of how cagey the subjects played it, the brain was always a step ahead. The decision had already been made before the deceptive subjects abruptly "decided" to act.

From these studies and others like them, it's clear: our subjective impression of "choosing" arises only *after* the decision has already been made as reflected in preceding brain activity outside our awareness (Chun, 2008).

But the story doesn't end here.

In further research on patients who remained conscious during a neurosurgical procedure, Libet identified a notable delay in pain perception. Only after the brain pain center had been stimulated for *at least half a second* did patients actually feel pain. Obviously, this is not how the sensation of pain works in everyday life. We don't put a hand on a hot stove without feeling it for half a second. Libet was finally forced to conclude that the brain must create a kind of "time warp." By projecting conscious experience backwards in time, the sense of delay is eliminated and the felt pain made to coincide with the actual event.

Libet assumed this same brain trick was at work in creating the conviction we have of conscious decision making. The actual brain decision is recast a split second afterwards in our subjective world as a conscious "choice" rather than what it is: the after-the-fact report of a decision already made. From the personal perspective it's like watching a delayed telecast and mistaking it for a live one.

In his book, *The River of Consciousness*, Oliver Sacks describes after-the-fact interpretations with respect to "myoclonic jerks" that sometimes occur as we are just about to fall asleep. These abrupt involuntary movements originate from primitive parts of the brain, but according to Sacks it's not unusual for them to be given special meaning a split-second afterwards by an improvised dream. Such dreams may be extremely vivid and have several scenes. "Subjectively, they appear to start before the jerk," Sacks explains, "and yet presumably the entire dream mechanism has been stimulated by the preconscious perception of the jerk. All of this elaborate restructuring of time occurs in a second or less" (Sacks, 2017).

Tor Norretranders refers to our after-the-fact reality as a *user illusion*. "We do not see what we sense," he explains. "We see what we think we sense. Our consciousness is presented with an interpretation, not the raw data. Long before this presentation, an unconscious processing has discarded information so that what we see is a simulation, a hypothesis, an interpretation; and we are not free to choose" (Norretranders, 1998).

One of Many: The Commonness of Human Illusions

The astonishing illusion—as jaw-dropping as it is—is only one of numerous illusions in the brain's bag of tricks. Take for example the optical variety. Gaze on the famed arch of St. Louis, a photograph or the real thing. Try as you may, your mind will not be able to "see" what is actually before you. Despite the upward, soaring impression of a slender arch extending skyward, in actuality, the width of the arch is precisely the same as its height. Your eyes and brain deceive you by collaborating in an optical illusion.

Time itself, seemingly so solid, is an illusion; the past, simply a memory that comes to us in the present and the future, a present imagination. Despite what our innate sense of time tells us, all actual human experience, all of it, is present.

Daniel Kahneman, Nobel-winning social psychologist and author of *Thinking, Fast and Slow,* considers cognitive illusions the result of hard-wired brain biases (Kahneman, 2011). Take for example, the almost universal *confirmation bias* where we tend to ignore or reject facts and observations that contradict an existing belief. It's why we see "hot and cold streaks" come and go instead of normal instances of regression to the mean. Closely related is the *coherence illusion* that results from our tendency to round off details so something is easier to understand and remember. In the process we often distort—sometimes in major ways—the actual facts. Kahneman also describes an *outcome bias* which leaves us vulnerable to unfounded generalizations. A friend has an unfortunate surgical experience. Without being aware of it, we fall into overgeneralizing this information so that it becomes a fixed belief that this particular kind of surgery is always risky. Illusions are a way of life for us.

One way to conceptualize cognitive illusions is to view them as paradoxes. Paradoxes are statements that seemingly cannot be true or false. Take the proposition, "I am lying." If true, I am telling the truth, even though I have declared I am lying. Such paradoxes arise at the intersection of two different planes of reference. This is the case with free will. According to the physical principles at work in the *material* world around us, it can't possibly be true, but in our *subjective* world of mind where imagination reigns, it is indisputably true. True and not true. Because we live in two different realities, the astounding illusion is a paradox.

Enter the Interpreter/Storyteller

Nassim Taleb, author of *The Black Swan: The Impact of the Highly Improbable*, views cognitive illusions as "narrative fallacies": stories fashioned without having all the facts to make sense of things (Taleb, 2007). The illusion of free will is one such fallacy: a deeply etched cognitive illusion designed to make coherent what otherwise would be unexplainable and even confusing. It is a summarizing narrative account, essential to our living subjectively surrounded by a material reality we have no way of knowing directly.

So, how does the brain go about creating this astonishing illusion? The work of Michael Gazzaniga, Director of the Cognitive Neuroscience Program at Dartmouth College, on "split-brain patients" provides a good starting point (Gazzaniga, 2011). For various medical conditions—especially treatment-resistant seizures—persons sometimes undergo neurosurgical cutting of a bridge in the brain (corpus callosum) connecting the two cortical hemispheres. In the case of a seizure patient, this procedure may be the only effective treatment to prevent seizures spreading from a small focus across the entire brain. Once the cortical hemispheres are separated, such spreading no longer occurs, but as a consequence, the two hemispheres are no longer able to "talk" with one another.

Gazzaniga made use of this unintended consequence. By taking advantage of being able to selectively present information to one side of the brain and not the other, he fashioned a fascinating series of studies. In one instance, the command, "Walk," was flashed on a screen so it was revealed only to the patient's right brain. Under ordinary circumstances a subject would stand up and walk towards the door and when asked why he did so, relate the message flashed on the screen. But with a patient's hemispheres no longer communicating, the response was quite different. (Recall the speech center is on the left side of the brain which in this situation was unaware of the command being seen only by the right side.) The patient found it difficult to reply, predictable since he was having to depend on his *unaware* left brain for any verbal response. In a bind, with no information he could respond to verbally, he was forced to *improvise* (confabulate) a response. "I'm going out to get a Coke," he hesitantly explained. It was all he could come up with.

Based on extensive research of this kind, Gazzaniga eventually concluded that during our waking life an area in the left side of the brain constantly spins an ongoing narrative that allows us to make sense of our world. He surmised there is a brain interpreter/storyteller module which in narrating a meaningful, ongoing story must constantly fill in informational gaps. He summed it up this way: "The interpreter's job is to tell a coherent story out of limited, after-the-fact information in such a way as to maintain the illusion of self-control" (Gazzaniga, 2011).

The interpreter module not only translates brain data into subjective material, it—or some other module in parallel—must also weave the translated information into an ongoing story at the heart of which is the self as chief protagonist. "We human beings have a centric view of the world," Gazzaniga contends. "We think our personal selves are directing the show most of the time... this is not true but simply appears to be true..."

This brain interpreter continuously supplies rationales for decisions already made. As William Wright, author of *Born That Way: Genes, Behavior, Personality*, puts it: "... we are all subjected to a continual barrage of biochemically delivered moods, nudges, shoves, impulses, cravings, and aversions [which] suggests that however bizarre, hurtful, or inappropriate one of our actions or mental responses may be, the interpreter module in our brain stands ready to provide it with a reasonable face" (Wright, 1998).

Well-known philosopher of mind Daniel Dennett has a similar take: "We are all virtuoso novelists, who find ourselves engaged in all sorts of behavior, and we always try to put the best 'faces' on it we can. We try to make all of our material cohere into a single good story. And the story is our autobiography. The chief fictional character at the centre of that autobiography is one's self." Dennett views the self as a "center of narrative gravity" (Minsky, 2006).

Fashioning the Story of Self

It's only in the interpreter/storyteller narrative that *meaning* appears. The situation is somewhat akin to what the Swiss psychologist, Carl Jung, called "synchronicity." Two persons travel from different cities halfway around the world for expressly different purposes. On arrival, they take lodgings in separate hotels for the evening. Early the next morning, each arises, bathes, dresses and embarks on a walk. Approaching a corner from different directions, suddenly, they are face to face. If they are strangers, no special meaning is ascribed to this brief encounter. Likely, they will pass and continue on their separate ways with little if any comment. The strictly determined material events leading up to their encounter have no special meaning other than to explain how the two arrived at the same point in time. But what if the two people share a past? What if their stories include being childhood friends, out of touch for decades? Given these subjective world happenings, this fully determined "meaningless" physical encounter emerges as an extraordinary storied coincidence. The encounter's *special meaning* is strictly a matter of interpretive storytelling.

In the guise of a first-person narrator, the interpreter/ storyteller selects from massive amounts of information. Incorporating billions of brain transactions and sensory inputs would be an invitation to informational disaster. Instead, with selective poetic license, the interpreter/storyteller spins a seamless and meaningful story absent the vast amount of details of goings-on in the material (brain) world. With actual causes and actions and decisions in the past and hidden, the creative fiction that unfolds fills the gap with the actions, thoughts, and feelings of a free-willing, choice-making protagonist at the center of a coherent and meaningful story.

Consider the experience of switching on a lamp. We flip a switch and, presto, the light comes on: a simple explanatory story void of all the intricacies involved. Nothing about neuronal electrochemical language being converted from neuronal impulses into muscle contraction powering the finger that turns on the switch. This is the wonder of a summarizing narrative spun by an interpreter/storyteller module in the brain. Although the intricate details of physical reality are sacrificed, the summary narrative story seems adequate enough for us to navigate the material reality around us.

In a more speculative vein, it's intriguing to consider the special challenge dreaming presents to the interpreter/ storyteller. Here with sleep accompanied by a drastic reduction in incoming brain data, the task of creating a coherent story likely becomes more daunting requiring as it does more "filling in" with creative riffs. The result: confusing, often bizarre scene jumping, clipped narratives we all know well which under the circumstances are probably the best the interpreter/storyteller can do, even on a good night. Limited information might also account for repetitive "stress" dreams—the same frantic scene time after time of being late or unprepared, unable to get where you need to be. Perhaps it's the best the interpreter/storyteller can do: repeat what was told before as a reflection of similar stresses from the past.

Demise of the Self as Action Figure

With respect to the age-old debate regarding free will, these neurobehavioral science findings and others put a substantial thumb on the scales as they convincingly rebut the long-held quaint belief in a ghost-like self inside our heads freely deciding and putting into action the way our lives go. The self as *free action agent* is seriously challenged. The neuroscientist, Antonio Damasio, sums it up: "There is no separate spectator for the movie-in-the brain. The idea of spectator is constructed within the movie, and no ghostly homunculus haunts the theatre" (Damasio, 1995).

Although we routinely assume cause and effect holds for the rest of the world, including our nearest relatives, other primates, we have insisted on drawing the line at human behavior. Humans are different, able to defy physical laws of cause and effect, *so we say*. But for how much longer? With the ascendancy of neurobehavioral science in the second half of the twentieth century, free will's dissolution is well underway. What have been portrayed as self-determined choices are now revealed as after-the-fact narrative accounts that emerge from an underlying biological material reality of which we have no direct knowledge. From this perspective the individual (self) is no more to blame for decisions and actions than would be a character in a favorite book.

Emergent World of Mind

Gilbert Ryle, the British philosopher who early in the 20th century labeled the idea of a mind separate from the brain as the myth of the ghost-in-the-machine, explained the illogic with a hypothetical story illustrative of emergence (Hooper, 1986). A visiting foreign student, still trying to master English, takes a tour of the Oxford University campus. Along the way his guide points out the library, the dormitories, academic buildings, research labs, and the chapel. At the end of the tour, appearing somewhat confused, the student reluctantly asks: "Thank you for pointing out all the different things, but where exactly is the University?" Ryle explained that the student's confusion resulted from a "categorical mistake" he was making when he failed to grasp that the university was not just another element, but a summative entity of greater complexity that emerged from all the rest.

Emergence is all around us. The core idea is this: something entirely different arises simply from a different *arrangement*. The same factors involved—nothing new added—giving rise to a higher complexity. Consider the two gases, oxygen and hydrogen. When brought together, water *emerges* with entirely different properties from oxygen or hydrogen alone. Similarly, from the combination of sodium (which by itself is a highly unstable metal given to exploding into flame) and chloride, a deadly poisonous gas, emerges a totally different substance: common salt.

In nature we see the same phenomenon in what are known as *murmurations*—the spectacular sky designs and images arising spontaneously from thousands of birds in flight. The patterns themselves are not caused or directed by some greater force. They simply emerge. (If you have never witnessed a starling murmuration, check out Neels Castillon's short Vimeo production, *A Bird Ballad*.) One bird researcher put it this way: "So one bird affects its seven closest neighbors and so on... This is how a flock is able to look like a twisting, morphing cloud with some parts moving in one direction and other parts moving in another direction and at a different speed" (Heimbuch, 2014). What from a distance appears to be orchestrated coordinated artistic moves is in fact an *emergent phenomenon* emanating from the independent action of thousands of birds doing their own thing without outside direction, design or additional cause.

Another fascinating illustration of emergence can be observed in the *Game of Life*, a board game created by the preeminent mathematician, John Conway. On a two-dimensional grid of identical square cells, each cell interacts with eight "touching" neighbors (up, down, to the sides, and diagonally). Each cell is always in one of two states: alive or dead. Any cell touching two or three live neighbors is alive; any cell with more than three live neighbors, dead. Adhering to a few basic rules of movement, live cells emerge undirected into fantastic action figures and stories (Dennett, 2003).

An example of emergence closer to our discussion of the interpreter's narrative story is *language* itself. From electrochemical processes in the brain, symbolic words emerge as nonmaterial building blocks of language in our subjective world with meanings totally different from the actions of neurons and neurotransmitters. You can explore brains down to the molecular level, you won't find language. Although it *emerges* from brains, it is not itself brain. The same can be said for the narrative self.

The emergence of language and the mind narrative qualify as what David Christian, a historian steeped in big history, calls a "threshold point! Language and our sense of self has moved life beyond simply surviving and reproducing to include a subjective world of meaning in the form of story" (Christian, 2018).

Origin and Future of Mind

Scientists and philosophers have no idea *how* the mind arises from the brain. Subjective reality from a material world, as Francis Crick discovered, is a good trick. Hard to figure. Indeed, it is an idea which when fully considered shocks the senses. Even scientists who have participated in the uncovering of the illusionary workings of the mind grow timid embracing it. Benjamin Libet, the researcher who demonstrated the "past tense" nature of choice, tried hard to preserve a small remnant of self-determination, claiming the self—although it could not initiate behavior—might possibly veto it! Brain scientist and Nobel laureate, John Eccles, insisted that brainwave signals construed to mean the brain has already decided on a course of action might be mere preparations for the actual self-choice itself. Such explanations seem strained at best. They suggest a desire to hold on at all costs to the image of conscious selves as free and in control. But trying to salvage free agency against a torrent of contradictory evidence gathering momentum all the time has the appearance of trying to swim up the proverbial stream.

What makes far more sense is to view the mind as an emergent phenomenon emanating from what the brain does naturally. While being tied together inextricably, the brain and mind are of entirely different natures, one material, one subjective. To confuse them would be similar to reducing the story of *Casablanca* to pixels and frames and expecting to find the meaning there.

To briefly recap what we have covered, although applying blame and punishment *fairly* has always been problematic, these practices remain entrenched due to a deep-seated belief in human free agency. The reasoning is straightforward. An essential aspect of our *homo grandiose* status is the unique ability to control our destinies through the exercise of willful choice; therefore, *unlike* any other life form, we are masters of our own fates, living outside the deterministic laws of cause and effect. This being the case we are necessarily blame and praise worthy. If in living our lives we excel, we merit special praise and recognition. In similar fashion, if we choose to break rules or laws, we deserve blame and punishment commensurate with the choices we have made. In relentless fashion neurobehavioral science is demolishing this portrayal of a free-wheeling self and replacing it with a shocking alternative: *the self as illusion*. Although the evidence is not in-the-bank conclusive, already it is compelling and far more persuasive than our belief in a free-willing genie inside our heads involved in choice making that routinely overrides the influence of genes and experience. If the "causative" self is only an illusionary character in a narrative account of what has already happened in strictly determined fashion and not by conscious choice, the basis for blame and punishment disappears, and we are left with *blameless responsibility*.

We see the sun pass over the moon, but we understand it's just an eclipse. We are stunned at watching an oversized harvest moon *shrink* while rising higher in the sky; no problem, we know it's an optical illusion. Similarly, there is no reason to think in time and with further clarification that the illusion of choice and self-determination will be any different: a compelling illusion we continue to live with but come to see for what it is.

Later we will pursue the implications of blameless responsibility for a radically different approach to criminal justice. But before we do, we go deeper into the workings of the self in a narrative world.

Chapter 8

Deep Story Telling and the Self

We're all stories in the end.
Steven Moffat

So, the mind emerges from the brain as the only reality we know. It's experienced by us as a comprehensible story with our conscious self as chief protagonist. This is who we are, characters derived from multiple genetic and experiential factors brought to narrative life by an interpreter/storyteller. Presumably, the resulting subjective narrative world we live in *emerges* from the workings of a vast number of material-based brain activities operating outside our awareness. Human experience arising out of a material reality becomes subjective and imaginative. This arrangement is consistent with what MIT professor, Marvin Minsky, has called the "society of mind" (Minsky, 1987; Singh, 2003). When Minsky describes the mind as emerging from the interaction of various brain agents it might well include an interpreter/storyteller.

Some will find this characterization highly disconcerting, but when compared to the alternatives—Crick's conclusion that we are "nothing more than a pack of neurons" or our own intuitive convictions of being free-willing genii randomly running our lives—mind as story is not really so weird or offensive.

Drilling Down on the Self

In other words, the life we live is derivative: a summarizing, *after-the-fact* narrative arising from brain events beyond our control or awareness that are fully determined, material, and non-narrative in nature. As central characters in this story we interact with other characters as we try to get what we want and avoid what we don't want (Goffman, 1959). Dan McAdams, a story psychologist, describes how a narrative context allows us to make sense of our past, understand the present, and look forward to the future with a sense of meaning and purpose. "Identity itself," he says, "takes the form of a story, complete with settings, scenes, character, plot, and theme" (McAdams, 2001).

Regardless of how it comes about, the character each of us experiences as self is a unique composite of a variety of traits including most importantly *conscious awareness*. Although other animals exhibit some degree of awareness, it appears much more prominent and nuanced in us. Hard to describe, yet so familiar this sense of being we have with its intermittent focused attention to matters at hand mixed with periods of drifting loose associations, imaginative leaps, and intruding memories. Our conscious experience comes to us in a collage of words, thoughts, images, sensations, and memories continuously shifting from "long shots" to "closeups" as our attention moves from diffuse to sharp. When attention is focused, aimless associations and memories are repelled only to seep back when it dims. Though conscious awareness continuously shifts from sharp focus to diffuse, always at its center is the anchoring self, the core of our awareness, experienced by us as a uniquely familiar and persisting entity through space, time, and experience.

As wakefulness gives way to sleep, awareness (including self-awareness) disappears on a nightly basis only to reemerge when we awaken with our life story and sense of self intact. Despite this daily sleep-wake cycle where periods of unconsciousness are standard fare, our stories retain a continuous, coherent quality. Each night when we dream we enter a different story—a story within a story—often bizarre, fragmentary, and disorienting, during which oddly enough our sense of self and our waking story remain intact as it does with age despite most of the cells in our body being replaced multiple times.

The Psychology of Self

The psychology of the self has long been a subject of study by philosophers and psychologists alike. Despite self and its story having now been revealed as a narrative illusion, many of these previous insights are still applicable in the same way the analysis of characters in literature, plays, and movies can be revealing.

Freud (like the ancient Greek philosopher, Plato) emphasized three working parts of the self: (1) an executive (Ego) constantly working to balance competing demands of (2) impulses and desires arising from the unconscious (Id) and (3) a conscience born of societal demands and later internalized (Superego). In this scheme the ego acts as chief executive officer in charge of overall planning and mediation between the disparate demands of the impulsive, emotional Id and the constraining, demanding Superego. As an establishing shot of the self, whatever individual differences it takes on, this general description remains relevant.

Though our story selves share a basic structure and certain universal traits, from one individual to the other there is tremendous variability. Consider the attributes of *self-awareness* and *openness to others* described in the conceptual tool known as the *Johari Window* (Luft, 1955). It's a diagram composed of four windows (squares), two at the top, two at the bottom. The top left window represents *public* information known *both* by others and the individual. The box next to it on top represents a *blind spot* of information known by others, but *unknown* to the person his or her self. Below, bottom window on the left contains *deceptive* or *façade-like* information known by one's self but unknown by others. Finally, the fourth window (bottom on the right) contains *truly unknown information* both to ourselves and others. The relative size of these various windows differs widely from self to self and results in vastly different characters that anchor the storylines emanating from the interpreter/storyteller.

Other notable differences in the self include the degree to which psychological defenses such as denial, projection, repression, and sublimation are used and reflected in different styles: naïve, paranoid, withdrawn, depressive, anxious, and adaptive. Personality specialists have also described variations of self-image (narcissistic/low self-esteem) and cognitive and emotional styles (extravert/introvert). Although the intricacies of the casting of the self remain largely unknown, these are all potential building blocks for what becomes the character of the self at the center of our life stories.

Perhaps the most notable characteristic of the self is one we have already considered: the universal conviction of *self-agency* that leads us to believe we freely choose our ways through life. Despite how much this conviction conflicts with how the physical world around us works, typically, belief in self-agency is unwavering. We hold to this quaint belief that we are controlling ourselves, unaware of the causative underlying brain activities arising from gene/environmental factors outside our awareness. This astonishing illusion is an essential element of who we are.

Likely this conviction goes unchallenged in large part because we are oblivious to the material operations of our brain and body that give rise to it. Take the experience of driving. Few of us have not had the sensation of suddenly *coming to* after miles of driving completely preoccupied with extraneous thoughts and memories. We pass it off as some kind of special multitasking ability rather than zeroing in on the fact we have gone for an extended period without being in control or even aware of driving. Unconscious driving is only the tip of the iceberg. None of the intricate workings of our brains and bodies ever appear on our conscious radar screens. Human experience is separate and apart, subjective not material, a story of its own.

Motivation is another distinguishing aspect of the character of self. The humanist psychologist Abraham Maslow described five levels of *motivation* from basic to highest: physiologic, safety, love/sex/belonging, esteem, and self-actualization (Maslow, 1987). Although to some degree all five are constantly in play, Maslow insisted that the lowest level dealing with significantly unmet basic life needs would dominate our awareness until satisfied in some minimal fashion. Only then would the self shift its attention in any major way to the next highest motivating need. Regardless of the unique trajectories of our individual stories, what motivates the self becomes an essential aspect of the evolving story. What Maslow had no way of knowing is that the self character—despite what it seems—is not the originator of the motivations that lead to various actions. These causative gene/environment interactions remain outside our awareness until—through the work of the interpreter/storyteller—they are portrayed in story form with the self as an action figure. In short, although we live with the unswerving conviction of self-determination, it's an illusion. Like a continuously delayed video, what we see, feel, choose, and do has already occurred and comes to us only in narrative translation (Suddendorf, 2018).

Many will consider the characterization of the self as an illusionary character in a story the greatest human put-down yet, but of course that's not all we are. The self is wrapped in a physical reality we perceive as our bodies—a surrounding context known only to us *indirectly* through our senses translated into brain signals and integrated into higher abstractions before they are finally offered up to us as the physical being we appear to be as story characters. This material human entity we experience subjectively has come to dominate the world.

This is not to say there are no disturbing aspects to life as after-the-fact narrative. One of the most profound is the sense of fatalism it threatens to engender. In fact it's why, at the brink of declaring human experience fully determined, many philosophers and theologians hesitate sensing to do so will lead to the loss of all restraint. What will keep persons from concluding they should throw all caution to the wind and do whatever excites or gratifies them at the moment? Why not? No blame, no restrain. At first glance the idea of the self as an after-the-fact illusion without actual control over what he or she does seems conducive to ultimate cynicism. But as we shall see in a moment this view of reality in no way suggests we make no choices.

Massive numbers of choices are made in our body/brain lives every second, 24-hours a day outside our awareness. The difference is they are not self-directed without cause, but strictly determined by an array of interacting genomic and experiential influences that determine all aspects of our lives—physical, social, psychological, and spiritual. In other words they are choices being made at another level as part of a material reality. But they are not choices made by the self and then when rule-breaking in nature blamed and punished by society.

Choice Redux

The spectacular history of human progress implies highly adaptive decision making. We live by choices, but not by the ones we think we make. What we perceive consciously as acts of self-agency are no more free choices than are the actions of characters in a novel. It is this illusionary sense of self-agency that gives rise to our twisted sense of the moral rightness of blame and punishment. With the relentless accumulation of supporting evidence, it seems likely we will eventually come to see our human experience the way we do magic. An evening in the presence of a world-class magician leaves us with vivid impressions of impossible happenings. We see and experience what we know cannot possibly be. A woman cut in half, an elephant disappear, or a death-defying escape. Thankfully we have the ability to intellectually override what appears to be unquestionably true. This is why we can enjoy and appreciate the illusions induced by skillful magicians. In similar fashion once evidence from the neurobehavioral sciences seeps into our social awareness, we should be able to accept the illusionary nature of self-agency as a reflection of something much more complex along with the far-reaching implications this realization holds for morals and the law.

I hasten to add, it's not that our narrative reality plays *no* part in decision making. It exerts influence, but only *indirectly*. Our self-responses to what we *experience in the story* become feedback to the brain and join with a myriad of other influences in determining future decisions and providing material from which the interpreter/storyteller can draw to weave its continuing story. The broader and more open our experience, the more potentially constructive it is for future reference, including necessary course corrections. In this sense, the self's assessments and reactions serve as an additional *sensory channel* to the brain. Perhaps it's this interaction between subjective and physical reality that underlies the development of so-called "psychosomatic" disorders.

Consider *psychological dwarfism*, a recognized form of shortened stature resulting from chaotic or abusive family environments. Here the psychosocial environment exerts a toxic influence on growth hormone production. When the person's social situation improves, the disrupted physiology corrects, and (if not too late) normal growth resumes (Ferholt, 1985). Similarly, in the area of allergies, full-blown, life-threatening asthmatic attacks can be precipitated by the mere image of an allergen (Rodriquez, 2015). Intense sneezing precipitated by imagined pollen in the air. In similar fashion, there's no reason to think more positive self-experiences cannot influence future brain actions. This may be the case with meditation. Numerous studies reveal changes in brain patterns brought on by the induction of relaxed focused attention (Lutz, 2008).

Neo-existentialism?

An after-the-fact narrative self—void of action—creates a challenge for certain philosophies. In 1942, when Viktor Frankl was Director of the Vienna Neurological Department of the Rothchild Hospital in Vienna, his wife was forced by the Nazis to have an abortion. Two years later both were imprisoned in the death camp at Auschwitz. Frankl was freed by Allied Forces in 1945 only to discover his wife, mother, and brother had died. Out of this horrendous experience he developed *logotherapy*, a philosophy aimed at finding meaning in life even in the harshest of circumstances (Frankl, 1950). Key to his approach was the assumption that despite whatever life throws at us, we retain our ability to choose. "Everything can be taken from a man but one thing: the last of the human freedoms—to choose one's attitude in any given set of circumstances, to choose one's own way." How would Frankl have dealt with evidence of an illusionary self?

Similarly, individual choice plays a central role in Jean-Paul Sartre's existential view of man's basic condition. "We are our choices," he declared, explaining how this was both a blessing and curse. The absolute freedom humans have to direct their lives, he insisted, also condemn us to blame and be blamed when our lives go astray. For Sartre *choice* was the key human attribute. We'll never know what his reaction would have been to neurobehavioral science's portrayal of choice as illusionary. It's difficult to see how he could have avoided a major revision in his philosophy, perhaps something along these lines: We are our choices... but we don't *make* our choices. Or, perhaps he would have concluded that the freedom we experience in our subjective realities is all that matters. Illusionary or not, he might say, be true to your life as you perceive it. Understand the story as meaningful.

In 1929 the ever-provocative Belgium artist René Magritte painted *The Treachery of Images*. Front and center is a smoking pipe, so realistically rendered you feel you can reach out and remove it from the painting. But just beneath the pipe appear the words: "Ceci n'est pas une pipe" or in English, "This is not a pipe." In similar fashion the experience of consciousness— despite what it seems—only provides a stand-in for a much more complicated material reality, one we can never know directly. Given our successful evolution, we have to presume this conscious *story map* mirrors what it represents close enough to provide useful guidance as we make our way in a world we know only through translation.

The Back Story: Strictly Determined

Of itself, a narrative recasting of brain phenomena need not diminish the value of human experience. "Mysteries do not lose their poetry when solved," says Oxford professor Richard Dawkins. "Quite the contrary, the solution often turns out more beautiful than the puzzle..." Many years ago, based on an interview with Albert Einstein, George Viereck wrote an article entitled, "What Life Means to Einstein" (Viereck, 1929). He quotes closing comments of the great cosmologist and physicist putting causation in perspective. "Everything is determined," Einstein said, "the beginning as well as the end, by forces over which we have no control. It is determined for the insect as well as for the star. Human beings, vegetables, or cosmic dust, we all dance to a mysterious tune intoned in the distance by an invisible player."

To be sure, the assumption that all things are strictly determined leaves much to be explained. Mysteries abound particularly with respect to the workings of the human brain and mind. We have only a rudimentary understanding of the narrative self; still, it's enough to put aside the fanciful idea of an unbound entity routinely guiding our lives in defiance of genetic and experiential influences. Just as a movie story has nothing to do with the mechanics and machinery of projection or the film in which it is embedded, the self story which emerges from a strictly determined material reality follows a different set of rules similar to that found in narrative fiction. The error our criminal justice system makes is to mistake the illusionary action of the self for the underlying determinative causes.

Human existence as *story* represents a new challenge to man's age-old claim of exceptionalism. Only a few months after his death, Ernest Becker, an American psychologist, was awarded a Pulitzer Prize for his book, *The Denial of Death* (Becker, 1974). Ahead of his time, Becker recognized the two entirely different worlds humans inhabit; one physical, the other, subjective— imaginative and symbolic. He viewed the search for meaning a central task of man's *symbolic* life. For Becker, imaginative creations were the ultimate triumph over death. Our stories, though they incorporate universals, are as unique as individual snowflakes. Although some seem to work out better than others, they are all unique productions. While death brings the end of a physical body, it completes a unique story which once lived is forever.

The most amazing thing about the imaginative stories we live is how they violate the laws of physics with ease. It's why we can move so easily from present to past to future. Why we can pursue the wildest and most fantastical thoughts. It's through such imagination that our personal stories intersect with those of others, often momentarily (strangers on the street, retail transactions, random crowds, etc.) at which point the entangled story ends until either we meet again *or* a memory is triggered with added associations that advance the story a bit further. After other intersections our storyline moves in parallel with that of others—some more intimate, most others less so—for more extended periods. Even fewer are those instances where our storylines become tightly entwined with others (relatives, friendships and love) as long as the stories last.

Self as Story Character

The self as story is presaged in Jostein Gaarder's best-selling 1991 novel, *Sophie's World*. Intrigued with philosophy, the book's 14-year-old heroine, Sophie, is surprised one day when she finds two cryptic questions in her mailbox: (1) "Who are you?" and (2) "Where does the world come from?" Soon after, she receives a series of written lectures on philosophy from someone she doesn't know. In time Sophie meets this mystery person, Alberto, who guides her through a review of Western philosophy, including contrasting ideas of the imaginary (idealism) and real (realism) which eventually helps the two of them solve the mystery they discover.

Alberto and Sophie conclude they are not who they seemed to be. At a garden party thrown by Sophie's mother to celebrate Sophie's birthday, Alberto reveals the startling details of their discovery. Strange beyond belief, they are all characters in a book in progress (titled *Sophie's World*) written by a Norwegian major, Albert Knag. The book is a birthday present for his daughter, Hilde, who is Sophie's age. Alberto and Sophie take this startling revelation as a challenge. In their world stripped of the illusion of self-determination, they work to uncover a different basis for finding meaning in life. (As we will see later, such a revelation also demands a reassessment of our ideas concerning justice and evil.)

Despite the narrative nature of the self, we are quite different from characters in books, movies, or plays. Our stories are not fixed from the beginning. We are not locked in like novels or screenplays. The subjective stories we live are constantly evolving as ongoing narrative translations of the material world changes.

Why the Illusion?

Why an illusionary sense of self-determination? Why would the brain undertake such an elaborate hoax? Apart from simplifying things, what could be the adaptive value? For starters, a sense of self-determination might well enhance *motivation*. In her book, *The Influential Mind*, Tali Sharot discusses intriguing studies that show a sense of being in control encourages behavior (Sharot, 2017). The question remains can motivation be retained when the illusionary nature of self-agency is revealed. Even though I know of no research on the subject, many persons through history who have questioned the idea of free will—even before neurobehavioral science drove the nail in the coffin—seem to have been well motivated.

A sense of self-efficacy also encourages positive emotions and the beneficial health effects linked to them. One imaginative study of elderly persons in a nursing home showed how a simple act of assuming responsibility (watering a plant) was correlated with greater happiness and increased social participation. Similarly, a greater feeling of self-control has been advanced as a cause of increased hope and less depression, stress and even premature death (Seligman, 2018).

A third possible role for the illusion of self-determination has to do with changing behavior. Years ago, the noted Stanford social psychologist, Leon Festinger, convincingly showed how our feelings, thoughts, and behavior are closely tied together in a finely integrated balancing act. When this balance is disturbed, we experience uneasiness—what Festinger called "cognitive dissonance" (Festinger, 1962). For example, assume a person dislikes his mayor's politics, but a close friend talks him into giving a supportive speech for the mayor's latest initiative. Based on dissonance theory, this public expression of support— in and of itself—is likely to result in the person feeling less oppositional to the mayor. Why? Because his negative *feeling* for the mayor now has to be squared with a contradictory *action* of support. Unconsciously the person accomplishes this with a positive shift in his view of the mayor.

But Festinger discovered an intriguing exception to

dissonance. It only develops if an action is undertaken *freely*. If a person feels coerced into doing something, no dissonance arises and there is no compensatory behavior change. Going back to our example, if someone *paid* the person to speak highly of the mayor, the shift to a more positive view would not occur. Perhaps in this same fashion a sense of self-determination instilled by the illusion of free will plays a key role in personal change. Only if you feel free are the rules of cognitive dissonance engaged.

When the Self Goes Away and Other Disruptions

After hours of unconsciousness, each morning when we awaken we still recognize ourselves, and despite all the ups and downs in our lives, unexpected turns, and numerous people entering and leaving, our self-awareness remains relatively unchanged and intimately familiar. We know who we are and what we have done and where we have been. We know our quirks. We know things about ourselves others don't. Our selves are truly anchors in our human experiences. If someone asks how old we *feel* (as opposed to what the calendar says), we have difficulty answering because we don't sense our inner selves as changing in major ways.

So it's shocking to hear reports of persons who abruptly lose familiarity with themselves. Apparently, the storyteller/interpreter giving rise to our sense of self can become dysfunctional. It happens sometimes with advanced dementia. Persons may forget where they are, what time it is, and who the persons are around them (even those intimately known for years). Much more rarely do people forget who they are, but it does occur.

The most dramatic instances of self-forgetting are abrupt instances of amnesia called *dissociative fugue*. In his *Great Courses* mind-body lecture series, Patrick Grim describes a famous instance of this condition from another century (Grim,

2017). On January 17, 1887 Reverend Ansel Bourne went to his bank in Providence, Rhode Island and withdrew $551 after which he disappeared. Later, upon his arrival in Norristown, Pennsylvania, he assumed the name A.J. Brown and set up a stationery and confection shop. Two months passed before he awakened one morning in a panic, completely unaware of who or where he was and without any recollection of what had happened. Eventually found by a nephew, he was returned to Providence and evaluated by the famous Harvard psychologist William James. Under hypnosis two separate persons—Bourne and Brown—appeared at separate times without knowledge of each other. (Reportedly, this strange case inspired the movie, *The Bourne Identity*, based on a 1980 novel by Robert Ludlum.) Nick Medford, a cognitive researcher at Brighton and Sussex Medical School, has identified the insula fold at the center of the brain as central to the perception that "this is me here and now." In a brain imaging study of 14 persons with depersonalization he found a notable absence of activity in this area (Thomson, 2018).

For the most part when storytelling and awareness of self is disrupted, it's only temporary. Cognitive degenerative diseases are a major exception. The mind requires an intact underlying brain. When it deteriorates due to disease, severe intoxication, or injury, the story and its chief protagonist go with it. But before the complete disruption of coherent story making, some individuals begin to have gaps that are the direct result of a failing memory. Because they cannot remember things they should know, they fill in with extraneous and often more exaggerated answers as best they can. Neurologists refer to this as *confabulation* (Hirstein, 2006). The noted sociobiologist Edward Wilson in his book, *The Meaning of Human Existence*, after acknowledging the self exists as chief protagonist in a brain-spun story, insists that such stories are totally "confabulated scenarios." (Wilson, 2014). He's mistaken. Central to confabulation is a failure of memory.

The normal storytelling at the core of human experience is not the result of faulty memory. It goes on in persons with perfect memory. It is a translation—not a confabulation—of brain language into narrative language in an attempt to give meaning and coherence.

No one really knows how or why an ultra-stable story creation (the self) suddenly vanishes and then reappears or moves from one "multiple" to the other. Similarly, it's not clear why in some persons who consume psychedelic substances such as psilocybin (mushrooms) there is a dramatic disruption of storytelling and break down of the boundaries between self and non-self to the point the self seems to completely disappear. In an intriguing *New York Times* article, "The New Science of Psychedelics," noted author Michael Pollan discusses a part of the brain affected by psychedelics known as the "default mode network (DMN)." He describes it as "… a network involved in a range of 'metacognitive' processes, including self-reflection, mental time travel, theory of mind (the ability to imagine mental states in others) and the generation of narratives about ourselves that help create a sense of having a stable self over time." Observation of persons in controlled settings who have received psilocybin and report the strongest sense of "ego-dissolution" (loss of sense of self) show the greatest suppression of brain activity in their DMNs (Sheldrake, 2020; Pollan, 2018).

There is much to be discovered about the narrative self and the story in which it's embedded. What we can assume is that they arise out of brain structures as an emergent phenomenon. Since the illusion of self is so central it would not be surprising to find an area specific to it such as the DMN. But for the most part how the deceptive brain conjures up the astonishing illusion remains unknown.

On the whole the interpreter/story-created self is incredibly stable as is the illusionary belief in its magical ability to choose and act outside the influence of genes and experience. Despite modern proclivities for criticizing the idea of mind and brain/ body as separate entities, that is exactly what we are left with. Two realities most certainly tied to one another—one emerging from the other—but without any substantial understanding of how this comes about. Donald Hoffman, author of *The Case Against Reality*, summarizes it this way: "What I'm claiming is that we're born with a virtual reality headset to simplify things to give us what we need to play the game of life, without knowing what the reality is... I think that everybody recognizes that we don't see all of reality. I'm saying we see *none* of reality" (Hoffman, 2019).

Mind as Virtual Reality

One of the most radical versions of alternative realities has us living in a *simulation* created by a superior intelligence. As outrageous as it sounds there are several Silicon Valley types and other scientists who believe chances are great that the lives we live—and even the Earth itself—are nothing more than a massive *simulation* created by a supercomputer built by a superior intelligence. According to a *New Yorker* profile of venture capitalist, Sam Altman, currently there are two high tech billionaires "secretly hiring scientists to work on breaking us out of the simulation" (*The Guardian*, 2016). In a similar vein Elon Musk has been quoted at a June 2018 conference: *"There's a billion to one chance we're living in base reality,"* which translated means it's a slam dunk the world we think of as real is actually a giant simulation. Richard Terrile, a scientist at NASA's Jet Propulsion Laboratory, brings the idea closer to home when he comments on rapidly evolving technology. "If one progresses at the current rate... a few decades into the future, very quickly we will be a society where there are artificial entities living in simulations that are much more abundant than human beings."

To be sure there are scientists of equal standing who think the whole "simulation hypothesis" rubbish. But much less so smaller scale *virtual reality* technology. It's worth reviewing for its possible implications with regard to the workings of the narrative self.

VR

Consider *Fortnite*, a recent video gaming sensation by Epic Games. Wildly popular it is now played around the world (125 million persons) "obsessively by children, rappers, professional athletes, and middle-aged accountants" (Pendleton, 2018). With a cartoonish swagger, a player's avatar (representative) enters a massive battlefield where players struggle against each other in a titanic war on a storm-battered island, all in 3-D, brilliant color, and stereophonic sound.

Recently, Epic Games staged its first *In Game* concert, featuring *avatar* Marshmello. It lasted ten minutes and was attended by 10-12 million global *avatars* who flocked to the stadium venue without leaving their homes. For this concert all weapons were decommissioned once players entered the virtual stadium. As an audience surprise, at one point participants suddenly found themselves becoming airborne after a zero gravity function was turned on. Participation in *Fortnite* is free. Epic Games makes its money by charging players for costumes and other accessories and perks using virtual in-game currency.

Another indicator of the growing popularity of virtual reality gaming is the success of touring video-game teams who now sell out stadium-sized venues. Spectators pay simply to watch giant screens where the teams are *projected* as their other-world, avatar characters battle each other. In 2014 *Amazon* bought *Twitch*, the world's most popular game streaming platform, for 1.1 billion dollars, its major audience these same spectators who pay to watch video gamers go at it (Sun, 2018).

In the late 1950s an early attempt at creating different realities mechanically was made by Mort Heilig. He called his somewhat clunky invention *Sensorama*. A person sat in front of a console and stuck his or her head in a hooded space where a film was viewed under unusual conditions. The viewer saw a stereoscopic color display backed up with stereophonic sound while fragrances from an odor emitter wafted through the air and the chair he was seated in vibrated. Heilig envisioned his machine as the future of movies (Rheingold, 1991). Although this was not to be, *Sensorama* did become an early precursor of *virtual reality*.

Present state-of-the-art VR allows a person fitted with a headset to experience a wrap-around alternative environment of sight and sound. Throw in special gloves (and perhaps even a sensor suit), add a joystick and the person is ready to interact fully with a fantastical world of simulated reality. For those readers who have never experienced VR, Jaron Lanier, a pioneer in the field, gives a taste of advanced VR in his book, *Dawn of the New Everything* (Lanier, 2017). "... it was natural to experiment with changing into animals, a splendid variety of creatures, or even into animate clouds. After you transform your body enough, you start to feel a most remarkable effect. Everything about you and your world can change, and yet you are still there... you can experience flying with friends, all of you transformed into glittering angels soaring above an alien planet encrusted with animate gold spires." Lanier goes on to explain how an immersion in VR can also involve *taking away* "... all the elements of experience piece by piece. You can take away the room and replace it with Seattle. Then take away your body and replace it with a giant body. All the pieces are gone and yet there you are, still experiencing what is left." Reading this you might well assume VR experience something akin to consuming a mind-wrecking hallucinogenic drug. To the contrary, in VR the mind and senses are perfectly clear. VR doesn't rely on distortions in brain functioning. It is an accurate reading of a newly simulated reality with the boundaries confronted in everyday reality shattered.

In a variant referred to as *augmented reality* digital imagery can be inserted into a real-world scene. For example, a digitalized version of a piece of furniture from a showroom can be manipulated in various positions in the room where it would be placed if purchased. There are rumblings in the tech community about a more nimble version of 3-D augmented virtual reality eventually replacing the smartphone which already has evolved into a mobile computer with a myriad of apps. Such a device presumably could be controlled by touching virtual reality buttons. Augmented virtual reality has been used for years in the construction of flight trainers for pilots.

The substantial difference between a movie experience and virtual reality resides in what is called the "framing effect." It's what separates the viewer's world from the movie world. The greater the distance, the less "real" the experience. The digital eye ware of VR greatly shortens the distance from the viewer and the simulated experience. By doing so the distinction between real and virtual is blurred. Currently, however, even using the most sophisticated VR technology, the VR experience remains distinguishable from our normal "reality." But experts anticipate that in the next generation or two, VR advances will narrow the gap further so that the difference becomes almost imperceptible.

Although descriptions of VR are difficult to render in translation, in trying to identify a best analogy for what neurobehavioral scientists are telling us about the emergent mind, I keep coming back to it. The Finnish scientist Antti Revonsuo reacts similarly when he says: "… we should take the concept of virtual reality as a metaphor for consciousness." He goes on to explain how the brain "… is actually creating the experience that I am directly present in a world outside my brain although the experience itself is brought about by neural systems buried inside the brain" (Revonsuo, 2016).

Howard Rheingold, author of the early best seller, *Virtual*

Reality, offers a similar understanding: "We build models of the world in our mind," he says, "using the data from our sense organs and the information processing capabilities of our brain. We habitually think of the world we see as 'out there,' but what we are seeing is really a mental model, a perceptual simulation that exists only in our brain" (Rheingold, 1991).

At present we can't know for sure, but the possibility of our subjective world being the world's best virtual reality—the product of millions of years of brain evolution—seems a reasonable speculation. If so, the self might well be viewed as a brain avatar and that should get your attention. In this arrangement brain decisions are projected onto an avatar figure in our subjective reality, the self. With this comes the capacity for consciousness: contemplation, introspection, subjective detection of sensory input, and reaction to actions taken. The avatar—its actions and responses—provides critical feedback to the brain for future decision making and storytelling.

Avatar or not, the rough outline of an illusionary self is in place. In the next chapter we take up its implications for the way we view evil. Traditionally, we locate evil in persons themselves, labeling them monsters, devils, and witches and, by doing so, justify vengeful punishments based on our belief that rule breakers do what they do out of intentional choice. Absent this assumption, evil becomes something entirely different and no longer can be located justifiably within the person him or herself. In fact to continue to do so would be immoral.

Chapter 9

Relocating Evil

The urge for retribution, therefore, seems to depend upon our not seeing the underlying causes of human behavior.
Sam Harris

The late Arizona Senator, John McCain, wrote an article for *The Wall Street Journal* entitled, "Vladimir Putin Is an Evil Man" (McCain, 2018). It's a well-written piece outlining nefarious Russian actions presumably directed by Mr. Putin. McCain viewed Putin as a threat to the liberal world order. But why "evil." He just as easily could have labeled the man "Russia centric," "exploitive," "ruthless," or "power hungry." Evil is its own special category, the ultimate in bad; but, it's hard to define precisely.

We Know It When We See It

The word elicits a strong gut response. Different from "wrong" or "bad" or "criminal." On closer scrutiny, it's used to mean two fundamentally different things; one, a product of nature destructively out-of-control, the other, moral depravity. *Natural evil* refers to horrific natural events: hurricanes, tornadoes, earthquakes, building collapses, and gas and chemical explosions associated with massive losses of life and property. The extent of destruction and how many lives are lost is what counts. In this variety of evil, intent or motivation are irrelevant. We don't blame nature for what it has done. In contrast, *moral evil*—the kind of evil we are concerned with here—implies a sentient perpetrator acting intentionally out of malice. The kind of person who for fun in a cold and calculating manner ties a man to the back of a vehicle because of the color of his skin and drags him through the streets until his head falls off. Moral evil concerns horrendous acts committed allegedly on purpose, by choice. Typically, this form of evil elicits feelings of outrage and calls for merciless retribution, even death. We know it when we see it, hear about it, or have it happen to us. If you polled the general population, likely you would find acts of moral evil highest on the scale of crimes deserving the death penalty. Whatever the bastards get it won't be enough.

Moral evil has certain core characteristics: (1) the infliction of extreme injury often in the form of maiming, torture or death; (2) blatant disregard for human dignity; (3) premeditated and calculated intent (fiendish); and (4) total lack of remorse. Despite being fully aware of what's being done and the horror it will cause, the perpetrator of evil chooses to do it anyway, at least that's the assumption. Although evil may be assumed to be encouraged by other factors (demons, devils, jinn, family, genes), tradition locates it squarely within the perpetrator and in so doing demands vengeful payback (retribution, sometimes even death itself).

But what we have covered in the last few chapters seriously undermines this characterization of evil. If evil actions are *strictly determined* by factors outside the perpetrator's subjective reality, beyond the influence or even knowledge (until after the fact) of the narrative self, locating evil within the person (self) is totally misdirected. Based on what we've reviewed, given an identical history, genetics, and situation any other person would have done the same. It's only our misguided insistence on viewing evil as a product of free choice that leads us to locate it within the person. But if this assumption is only an illusion; if the self's role is an after-the-fact narrative recounting of previously-made brain decisions, locating evil in the person is unjustified.

Disturbing Implications

At the sentencing of Theodore Kaczynski (the Unabomber), Susan Mosser described the terrible mix of rage and grief she felt in 1994 as she watched her husband's life slowly ebb away on their kitchen floor, the victim of a fiendishly constructed package bomb. She sobbed as she described her young daughter watching the life of her father's razor-blade-and-nail-riddled body ebb away. "My children are bleeding from their souls," Mrs. Mosser told the sentencing judge. "Please keep this creature out of society forever. Bury him so far down he'll be closer to hell, because that's where the devil belongs."

Some evil appears so monstrous, killing the perpetrator seems the only adequate response. Ultimate punishment for ultimate evil. Even the most strident opponent of capital punishment, finding herself in Susan Mosser's place—grieving the loss of a loved one savagely murdered—might well have similar feelings and think the perpetrator's death the only just outcome.

Even so, neurobehavioral evidence shows us such feelings are misdirected. It makes the case for a radically revised version of evil; one that relocates it *outside the self* in the causal determinants themselves—genes and experience. This makes for an unsettling conclusion: *no person can ever justly be considered evil*. Responsible, yes, but not evil. It's the underlying causes and the horrific acts they bring about that deserve the label, not the perpetrators themselves (narrative selves). Given what now seems more likely than not—the complete inability of the self to choose or act—the only reasonable basis for labeling persons evil would be to construe them as *natural evils* similar to powerful tsunamis or highly destructive tornadoes acting without intention. Even so, because they live in a social world, they still must be held responsible. Blameless, but responsible, subject to appropriate consequences—sometimes extreme—but absent punishment.

Some have argued retribution (punishment), apart from

being vengeful payback, is essential for another reason: to allow grieving survivors emotional resolution. It's one thing to excuse blame for minor rule breaking and even more serious violent crimes but an entirely different matter to do so for the most heinous acts of killing and maiming. Surely, at some point vengeance becomes a right of the aggrieved.

I admit, first considered, the idea of "excusing evil" is deeply troubling. And for awhile the more you think about it the more disturbing and outrageous it seems. Take the Tsamaev brothers in Boston. They constructed two homemade pressure-cooker bombs and positioned them several hundred yards apart near the finish line of the 2013 Boston Marathon. Detonated 12 seconds apart, the two explosions killed three persons and injured hundreds more. Sixteen persons lost limbs. Tamerlan died in an attempted escape, but his brother, Dzhokhar, lived to be convicted of multiple charges, including use of a weapon of mass destruction, for which he was sentenced to death. Not evil?

Reportedly an avid gambler and germaphobe, Las Vegas shooter Stephen Paddock was overheard cracking jokes with the hotel staff as they unwittingly carried bags of weapons to his room where he leisurely went about his way making preparations for a cold-blooded mass killing. Following a week of gambling and drinking, Paddock finally got down to the real business he had come for as he directed automatic-like gunfire from his 32nd floor window onto a crowd of 22,000 at a 2017 Las Vegas country music festival. Before finally taking his own life, he had randomly killed 58 persons and wounded another 400 in the deadliest mass shooting by a lone gunman in U.S. history. Hard to see as anything but evil.

Our traditional sense of evil would definitely include serial killers. Gary Leon Ridgway was one of the more prolific. Over several decades the so-called Green River Killer wantonly murdered until he was finally apprehended and convicted of killing 49 women and young girls. Most of his victims he raped and strangled to death before dumping their bodies in a remote forested area where periodically he returned to have sex with what remained. Beyond blame? Undeserving of punishment?

"Evil" immediately comes to mind when we consider the seemingly endless school shootings our country is experiencing. Loss of life was greatest on April 16, 2007 in Blacksburg, Virginia when undergraduate, Seung-Hui Cho, using two semiautomatic handguns, killed two persons in a dorm and then, after chaining shut the three main doors of an academic hall, murdered another 30 helpless victims. Is retribution really not appropriate? Our gut tells us otherwise.

Over the past year hundreds of instances of sexual abuse have come to light, many involving well-known personalities. The extent of this ongoing widespread assault has now been acknowledged and corrective actions including arrest, demotions, and public disgrace have begun. Such abuse has prompted the use of the term "evil" being applied to non-killers. Larry Nassar was a 54-year-old Michigan State University and USA Gymnastics physician. For over two decades, under the guise of providing expert medical physical therapy to women athletes, he sexually abused more than 350 women. At one of his court hearings 150 women confronted him face to face and with mixtures of rage and hurt labeled him a vile, monstrous, disgusting creature. Presiding over his sentencing hearing, Judge Rosemarie Aquilina said she was "honored" to sentence Nassar "to die in prison" and suggested he deserved being sexually assaulted himself. If not evil, what do we call it?

On the morning of October 27, 2018 Robert Bowers entered the *Tree of Life Synagogue* in the Squirrel Hill section of Pittsburg. Armed with an AR-15 rifle and three handguns, he shot worshippers indiscriminately, shouting: "All Jews must die." Twenty minutes later 11 persons were dead and six others seriously wounded. The victims included two brothers, a man and woman who 60 years earlier had been married in the *Tree of Life Synagogue* where they were gunned down, and a 97-year-old woman who had survived the Holocaust only to die now in her home synagogue years later. As Bowers tried to escape the synagogue, he was shot multiple times by a Pittsburg policeman and taken into custody. When asked later why he had committed the massacre, he said: "I just wanted to kill Jews." In the midst of grief over this horrific crime there were vengeful calls for the death penalty. Justified?

Our sense of evil is not restricted to crimes involving the killing of large numbers of innocent people with warlike weapons or extreme sexual abuse. Sometimes it arises due to the innocence of the victims. In February 2019, investigators interviewed Christopher Watts in a Wisconsin prison (Swanson, 2019). Over five hours he related how when his pregnant wife confronted him with an affair he was having and threatened to leave him with their two younger children, Celeste, age 3 and Bella, age 4, he grabbed her around the neck and strangled her to death. The noise woke both children. When they asked what was wrong, Watts told them their "mommy didn't feel good" as he wrapped his dead wife's body in a sheet and clumsily carried it down the stairs and dragged it out to his truck. Then with his two young children sitting on a bench seat in the back, he drove to an oil field about an hour away where he disposed of his wife's body. When he returned to the truck, the youngest daughter Celeste was first. He put a blanket over her head and strangled her as her four-year-old sister looked on. After dropping the child's dead body in an oil tank, he returned to the truck where according to him, Bella asked, "What happened to Cece? Is the same thing gonna happen to me...?" Watts told the investigators he wasn't certain what he answered. He just remembered Bella's last words: "Daddy, no!"

How can this not be evil? The answer is *it is evil* as are all of these other instances, but the evil doesn't reside in the perpetrator where self-directed intent and choice are only illusionary after-the-fact accounts. Seen in this light, the assignment of evil to the individual is a modern version of scapegoating. *It's the acts themselves and their true causes that are evil.* If actions are strictly determined outside our awareness, it is immoral to apply the term "evil" to the perpetrators themselves, even a man like Christopher Watts. They are victims as well. It goes against every moral fiber of our being, but it is a direct implication of what neurobehavioral science is telling us.

Evil Endures But in a Different Place

Despite how they are traditionally portrayed, evil acts are the product of ongoing gene/experience interactions, all outside the awareness and control of the narrative self. Even so, I hasten to add holding such rule breakers *responsible* without blame implies doing *whatever* is necessary to protect the public. Accordingly, those persons I have named in this chapter, judged as blameless but responsible, might well have been contained for the rest of their lives—not as an act of vengeful punishment, but as a matter of public safety.

Horrific killing, torture, violence, and extreme abuse, tragically, they all exist in abundance. But given what we have reviewed, human "evil" is more appropriately characterized as the *natural variety* similar to famines and plagues as opposed to being the immoral work of evil persons. As with natural disasters, serious rule breaking must be protected against and prevented as much as possible, but blaming only serves to divert attention from the true causes. Human criminals and hurricanes—neither is to blame, but both are to be prepared for and dealt with as smartly as we can which leads us to the next chapter: *Blameless Justice.*

Chapter 10

Blameless Justice

Free will is an illusion, but you're still responsible for your actions.
Michael Gazzaniga

In his speculative novel, *Erewhon,* 19th century novelist and painter Samuel Butler lampooned the idea of blame by portraying an upside-down fictional country where sick persons are prosecuted and punished while criminals are diagnosed and treated! Way ahead of his time, Butler reasoned since all behavior was strictly determined who gets blamed was more lottery-like than a matter of justice. Why treat the ill and punish those who break the law? Both are rule breakers, neither having chosen what has happened. At the time he wrote *Erewhon,* Butler's views lacked scientific backing; but, as we have seen, subsequent discoveries greatly strengthen his case. What Butler was speculating about is now established. The self that we blame and punish is an illusion for which blame assignment is at best wildly arbitrary, and at worst, immoral.

By now it should be clear, blaming is a misguided shortcut. Far easier to scapegoat "criminals" than to tease out various causative factors such as genetics, poverty, racism, severe abuse, addiction, and mental illness. Much simpler to incarcerate and punish than to offer innovative programs aimed at both protecting the public and providing true rehabilitative education and training. And as long as a misguided belief in free will and self-determination remains a foundational assumption of "criminal justice" this misdirected *blame game* will continue as part of a fantasy moral order that vengefully punishes or sacrifices rule breakers. But as contradictory neurobehavioral evidence continues to accumulate, this assumption likely will look less and less credible. As neuroscientist, David Eagleman, author of *Incognito: The Secret Lives of the Brain*, puts it: "Blameworthiness is a backward-looking concept that demands the impossible task of untangling the hopelessly complex web of genetics and environment" (Eagleman, 2011). If there is a moral judgment to be made, it's about the actions themselves and their underlying causes, *not* the narrative self portrayed in the mind story. Once this proposition is accepted, our entire criminal justice system goes wobbly as the similarities between modern practices of blame/punishment and ancient sacrificial rites of atonement become more and more obvious.

Once Again: Self Reconsidered

To briefly summarize, based on the best evidence available, "free will" is an illusion conjured up by a deceptive brain; human experience a narrative shorthand rendered as a *stand-in* for a much more complex reality, greatly simplified but close enough to provide a useful guide for survival in a world we never can know directly. Call it *primary* virtual reality. Central to this narrative is a chief protagonist experienced as "I" and "me." In this story role we have the compelling sense that we are deciding, acting, and controlling when in fact it's all an after-the-fact illusion. The bottom line is this: blame is assigned based on pseudo-choices made by a storybook character. Revealed for what it is, this is like prosecuting an avatar who "chooses" to shoot down an opponent in a virtual reality game or handing out a death sentence to the actor Anthony Hopkins for primitive killings committed by Dr. Hannibal Lecter in *The Silence of the Lambs*.

Some will consider this view little more than reductionistic nonsense. But as Nobel laureate Steven Weinberg cautions in *Dreams of a Final Theory*, "... the opponents of reductionism... are appalled by what they feel to be the bleakness of modern science. To whatever extent they and their world can be reduced to a matter of particles and their interactions, they feel diminished by that knowledge... I would not try to answer these critics with a pep talk about the beauties of modern science. The reductionist worldview is chilling and impersonal. It has to be accepted as it is, not because we like it, but because that is the way the world works" (Weinberg, 1994).

Despite revelations about quantum world spookiness, in the macro material world random acting agents *willfully* pushing aside all other influences whenever they choose are not to be found. Although many details remain to be worked out, the evidence for this view is substantial enough so that the burden of proof now falls on those who claim willful choice the cause of rule breaking and the justification for vengeful punishment.

But important to keep in mind, a deterministic perspective should not be allowed to obscure an extraordinary aspect of human experience: the capacity of the emergent interpreter/ storyteller to give rise to an imaginative narrative world that soars freely above the deterministic cold mechanics of a material reality.

Myth of Punishment as Deterrence

There will be those who balk at the thought of banishing punishment. Blame or no blame, they will say, punishment must be preserved if for nothing else the deterrence value it carries. If it works to reduce criminality, it should stay as a necessary cost of doing the business of criminal justice. A reasonable idea, but evidence doesn't support it.

When put to the test, punishment has little deterrent effect. Going back ten years, a 1999 meta-analysis of 50 separate studies looked at 336,052 California offenders to test the following proposition: *the more severe the punishment the less likely an offender will be to re-offend*. Controlling for risk factors such as criminal history and addiction, the investigators drilled down on the connection between *amount of prison time* and *repeat criminal activity*. What they found surprised them. Those persons imprisoned for 30 months or more returned at a rate of 29% compared to 26% of those spending 13 months or less in prison. Rather than being a deterrent, prison proved to be an enabler. The more prison time the more likely a return visit to prison. More than double the amount of prison time was associated not with lesser risk but with 3% *greater* risk of return. Also of note, serving time inside a prison produced a 7% higher rate of recidivism compared to a community-based program.

A more recent review (from the *Sentencing Project*) reached the same conclusion: "Existing evidence does not support any significant public safety benefit from imposing longer prison terms. In fact, research findings imply that lengthy prison terms are counter productive." The author, Valerie Wright, characterizes the *Three Strikes* law (three criminal offenses and you stay in prison for life) as well as the practice of handing out "mandatory minimum sentences" unnecessarily costly and without any demonstrated public safety benefit (Wright, 2010).

In retrospect the revered Chief Justice Oliver Wendell Holmes appears to have gotten it wrong when many years ago he explained his view of deterrence. "If I were having a philosophical talk with a man I was going to have hanged, or executed," he asserted, "I should say, 'I don't doubt that your act was inevitable for you, but to make it more avoidable by others we propose to sacrifice you to the common good. You may regard yourself as a soldier dying for your country if you like. But the law must keep its promises'" (Pinker, 2002). We know better now. Despite its vaunted reputation, punishment is a poor deterrent.

Others will interpret the idea of *responsibility-without-blame* as just another liberal ruse for being soft on crime. To the contrary, as the case against free will strengthens, whether you are conservative or liberal, setting aside blame becomes nothing short of a moral imperative. Still, even a blameless society cannot tolerate perpetual rule breakers, especially when the rule breaking involves violence or major crimes. In some instances, containment—sometimes even for life—will be necessary. Even so, this will not require mass incarceration settings, punishment for the sake of punishment, and tactics of persistent disrespect and demoralization.

Time for Radical Change

For several decades efforts in prisons have been sidelined. In their book, *From Retribution to Public Safety*, William Kelly and his colleagues describe a "tough on inmates" philosophy prevailing since the early 1970s (Kelly, 2017). It started with a questionable series of studies casting doubt on the usefulness of prison programs (Martinson, 1974). The results were snatched up by policy types, and in a short while programs to educate and job train were shown the door. Even though the majority of prisoners were serving time for nonviolent crimes, punishment behind walls became the main focus. What was left were overcrowded dangerous incubators of criminality, racist conflict, violence, and recidivism. Ironically, a main contributor to rising prison/jail populations—addiction—for the most part went untreated and even worsened in prison confines, and mentally disordered persons increasingly found themselves criminalized in lieu of treatment.

Adding to this bleak picture was a growing criminal justice tilt toward longer sentences (automatic minimal sentences being a part of this trend). A recent *New York Times* editorial— "New York Forgets Its Juvenile Lifers"—describes how things went bad for 17-year-old Carlos Flores when he and three others attempted to rob a bar in 1981. Even though it was one of the accomplices who shot and killed a police officer, Flores was convicted of second-degree murder and given a 21-to-life sentence. More than 37 years later the New York State Parole Board was still insisting Flores' release "would not be compatible with the welfare of society..." This was 25 years after Flores' last write-up for a disciplinary infraction because of his *being too slow in the mess line.* A recent evaluation of Flores declared him "the lowest risk of danger to the public." One editorial writer expressed the obvious: "Mr. Flores is being locked up not because he's a threat to society or has failed to show that he's changed, but for a crime he committed nearly four decades ago."

Unfortunately, Mr. Flores is one of many. A 2016 American Civil Liberties Union review found 12 states with prisoners serving 40 years or more for crimes committed as juveniles! Criminal justice blame-and-punishment philosophy unnecessarily brutalizes lives.

How Other Countries Handle Criminals

I've had a number of conversations with knowledgeable persons about criminal justice redesign. Typically, there is agreement on the major deficits of what we currently have. Too much regimentation. Poor living conditions. Too little rehabilitation. But beyond this, disagreements break out all over. After all, the people we are dealing with are criminals, many of them too dangerous for anything other than highest security isolation. At this point when I inject some of what other countries have done, typically, I am met with stone-wall skepticism. Persons ask if we are talking about "the same kind of prisoners."

Usually, I start with Norway where the motto of the Norwegian Correctional Service is: "Better out than in." Since 1998, when the program underwent a complete overhaul, the main correctional emphasis has been on job training, therapy, and basic education. Recidivism is now less than 30%, the lowest of any European country. Most Norwegian prisoners engage in work, much of it facility farming or various repair activities. The settings are "open." Prisoners are housed in private units reasonably well outfitted with television sets, mini-refrigerators and private bathrooms. They are permitted and even encouraged to engage in recreation such as swimming, tennis, and other sports. A good example of this open approach is the Halden Maximum Security Prison located on 75 acres surrounded by blueberry woods close to the Swedish border. Its main focus is rehabilitation aimed at preventing prisoners from ever returning.

But prison reform is not restricted to Norway. In early 2013 a delegation of U.S. corrections and criminal justice system leaders visited facilities in Germany and the Netherlands. They were impressed with the extensive use of fines and community service as *alternatives* to incarceration, noting how despite shorter sentences there were lower rates of recidivism. As in Norway, emphasis was put on work and job training as opposed to punishment. The delegation was impressed with the respect given prisoners in the form of increased privacy and service. They also noted how prisoners in both countries retained their right to vote. (In most states in the U.S. prisoners are stripped of this right.) Finally, mention was made of the extremely rare use of solitary confinement (Erbentraut, 2017).

Criminal Justice in a New Key

In a *TIME* magazine article, entitled, "The Age of Innocence," author Robert Wright described the horrific murder of first grader, Kayla Rolland. The fatal shooting took place the day after two children had quarreled. At school, Rolland's classmate, a six-year-old boy, pulled a .32 semiautomatic pistol from his trousers and said—"I don't like you"—before shooting the girl point-blank in the chest and killing her. Wright provided an unusual perspective. "The more you know about what makes people bad," he said, "the more you realize all murders are double tragedies." An end to blame-based justice will bring a view of rule breakers as *victims* themselves. Eventually, decline in the blame myth opens the door to a different perception of wrongdoers and an end to the rigid distinction between perpetrator and victim.

There's no need to start from scratch reforming mass incarceration. As we have seen, the forward work has already been done in other countries. For starters, high security need *not* require human deprivation, punitive isolation, or continuously dangerous living conditions. We have become so accustomed to current prison/jail culture we dismiss as unworkable viable alternatives. With advances in electronic surveillance and modern management, there is no need to employ paramilitary operations as a standard approach to wrongdoers, even the most violent. Aggressive measures should only be utilized when essential to security and public safety, and even then there are no reasons why high security settings should ever include extended periods of isolation.

Many of today's prisoners could be managed on the outside with ankle bracelet monitoring. The same for those who may require a period of containment and then be released. Through advances in satellite detection and tracking, parolees can be monitored 24-hours a day if necessary. Some might have tiny microchips attached to their bodies serving as pagers. Using voice-recognition technology, parole officers could determine location and status of a person at any time. Through satellite links such devices are capable of triggering an alarm if a person approaches a place or person considered "off limits."

Limiting containment only to those who cannot be handled any other way has implications far beyond the rule breakers themselves. We can start with the cost savings. Mass incarceration is an expensive proposition. This is truly one of those situations where the top-to-bottom redesign needed might well cost less.

Additionally, the negative effects of standard prisons are not limited to the inmates. Their families are at great disadvantage as reflected in elevated rates of mental and physical health problems as well as addictions. Adolescent boys with incarcerated mothers are 25 times more likely to drop out of school which increases their own risk of future criminal activity and incarceration. Former Attorney General Loretta Lynch described the multiplier effect of incarceration this way: "Put simply, we know that when we incarcerate women we often are truly incarcerating a family in terms of the far-reaching effect on her children, her community, and her entire family network" (Stillman, 2018).

Excessive cost, ineffectiveness, and negative consequences are enough reason to move away from prevailing prison practices. But from what we have reviewed in this book, the more profound reason is because they are immoral. Instituting a blameless approach will require seismic changes in criminal justice throughout; not just prisons. But prison redoes will be a good place to start.

Current emphasis on punishment, intimidation, and disrespect needs to be scrapped. It should be replaced by respect, expectations of regular social norms, and a priority commitment to rehabilitation. Overly regimented institutions of mass incarceration will be redesigned to be more like residential academic campuses. Policies regarding major goals and ways of achieving them will be completely rethought. But such changes will only go so far in the absence of a new culture and recruitment of personnel who reflect it. Those individuals who see their work as mainly enforcement of punishment and strict discipline need not apply. Many of those currently in these jobs likely would not be chosen.

Some will suggest *blameless* settings will be so appealing, persons on the outside will be encouraged to commit crimes just to get in. I think not. Even blameless constraint—inside or outside a facility—restricts a person's ability to move about and live life on his or her terms. It represents a considerable disruption to one's life.

A shift from control and discipline to intensive education,

training, and socialization will require a major change in forensic culture. Ultimately, the selection of different personnel who support such changes will be determinative. Structurally, the gray, stark prison appearance will give way to more normalized colors and a non-penal design. Many years ago I had the opportunity to work with the warden of one of California's high security prisons. One of few women wardens, she was unusual in many ways. She had come up from the ranks, all the way from being a secretary! A soft-spoken attractive woman, in her late fifties, she surprised people with her toughness when the occasion called for it, but she also showed a deep concern for prisoners, personally interacting with them on a regular basis. They liked and respected her. She hated the prison environment—the culture as well as the bleak setting— but there was little she could do about it, except for one thing: she insisted the outside grounds of the prison be filled with gardens and flowers. The more roses, the better. Unlike what most visitors—relatives and officials—find when they drive up to the outside of a prison, visitors to her prison were met with colors and greenery and perfectly-manicured grounds, courtesy of dedicated prisoners. A small thing, but it gave the institution a decidedly different feel. Settings matter. Current prison structures designed to emphasize dehumanized control and punishment are incompatible with true rehabilitation.

Often portrayed as beyond the understanding of professionals, experts, and executives in other fields, most prisons are directed by "insiders" relatively immune to change. As a result there has been a surprising lack of creative oversight. Imagine the possibilities of having executives from other fields—the likes of Jeff Bezos, Meg Whitman, Satya Nadella, Mary Barra, Jamie Dimon, Bill Gates, and Oprah Winfrey— brainstorming alternative goals, strategies, and methods for a *blameless-but-responsible* approach to criminal justice. I suspect their suggestions might well include replacing career prison management with open-minded executives who have demonstrated experience turning around troubled institutions.

Judges and Juries

A blameless approach will require major changes in the law itself and in the ways judges and courts conduct business. For all indicted cases of rule breaking, the first phase of a *blameless trial* will focus exclusively on whether or not the person committed the crime, but it would *not* involve the current circus-like antics by both defense and prosecution aimed at establishing the defendant's state of—guilty or not guilty—mind at the time of the crime. Such speculations as to degrees of a "guilty mind" which invariably lead to conflicting opinions would have no place in the first phase of a blameless trial. The only question under consideration would be: *Did the person do it?*

If the answer is yes, the second and far more complicated phase of a blameless trial would begin. Forced to put aside concerns with degrees of guilt and required punishment, the court will focus primarily on two matters: (1) *Requirements for public safety*. A judge and jury (with the help of specialists) would undertake a thorough assessment of the person's risk for violence, including among other things an in-depth evaluation of the nature of the crime, previous history of violence, level of remorse, and markers of future dangerousness. Based on these evaluations a decision would be made as to the initial level of containment—if any—which would be required. For most nonviolent persons, these requirements would be minimal, but for the most violent, the highest level of security needed to insure public safety would be established. This initial decision regarding violence risk would be reassessed at regular intervals. (2) *Remediation needs*. In a blameless criminal justice system, rehabilitation will no longer be an afterthought. For the majority of rule breakers it will be the central focus. This radical shift will require more detailed understanding of the rule breaker's previous history of rule breaking, personal interests, abilities, motivations, predisposing factors such as addiction and mental health problems, and his or her educational status and overall level of competency in order to come up with a preliminary plan for optimal rehabilitation and an estimated time requirement. Such an effort on the court's part will depend extensively on a selective mix of behavioral and job training experts, criminologists, medical specialists, psychologists, psychiatrists, and addiction specialists.

Our current system's one-size-fits-all approach will be replaced by an emphasis on individual differences. Rehabilitation efforts will be far more detailed than what now exists. An individual rehab plan might well include medical treatment, counseling, addiction therapy, and focused education and skills training—all directed toward helping the person achieve a reduced risk of future criminal rule breaking and a more stable and satisfactory life.

Reforming Rule Breakers

Although abuse of drugs and alcohol are major causes of criminal activity, other than failed attempts to control substances in prisons, these problems (with the exception of a few special courts) are largely neglected. Prison drug and alcohol use remains widespread, supplied by a sizeable black market. I have been told by a number of inmates if you have something to trade, whatever drug you want you can get. In a blameless criminal setting the treatment of drug and alcohol addiction would be seriously undertaken with state of the art interventions and well-trained personnel. If a person's crime clearly arises from substance use, a blameless court may well steer him or her directly into long-term rehabilitation that includes job training and ultimate job placement. In the current penal system not nearly enough drug and alcohol diversion occurs and often the most effective treatment methods typically are not available.

The same is true of mental illness. Far too many homeless persons with mental problems end up jailed or sent to prison (Torrey, 2010). Although lip service is paid to specialized services and treatment, usually this is an extremely marginal part of the inmate's total penal experience while little is done to protect these persons from prison predators. Mentally disordered offenders remain some of the most vulnerable persons in prisons. In a blameless system, the individualized treatment of these disorders by experienced personnel would be a major priority undertaken in treatment settings, not prisons. As it is in our current criminal justice system such treatment is either completely ignored or given only lip service.

With punishment eliminated as an objective, a blameless approach will do away with *fixed sentences* and replace them with time estimates (subject to change, depending on the person's progress) of how long it will take to significantly improve a person's chances of avoiding wrongdoing. In the absence of a history of violence, many persons will not require *any* incarceration. Instead, they will be diverted to alternative rehabilitative and education/training programs. A similar approach will apply to persons who, although initially assessed as dangerous, after critical program progress are judged safe and able to continue without incarceration. As it is, our current prisons and jails are run as though *most* inmates are violent and always will be. This is a myth.

Violent Rule Breakers and Blameless Justice

Those rule breakers judged high risk for violence would be sent to special containment settings with personnel skilled in working with such persons. The rehabilitative program would be designed especially with possible violence constantly in mind, but it would not allow extended isolation and would be specialized in using the least amount of force necessary when occasion demanded it. Such programs would be designed to insure public and staff safety while at the same time treating the person with respect and dignity. If and when the time comes that a person is judged no longer a danger, a more detailed program of rehabilitation would be implemented. For those thought to be an ongoing violence risk, a humane program of extended containment—sometimes lifelong—may be indicated, but without punishment.

Persons who commit violence vary greatly in their risk for future violence. For example, consider the following list of hypothetical murderers, each with a different life story that brought him or her to the killing moment. They illustrate the diversity of cases that might come before a judge and jury.

Without any previous criminal history, in a fit of jealous rage, a woman professor knifes to death her long-time lover.

An intoxicated teenager runs a red light and crashes his car into a minivan, killing a family of four.

Looking for his next fix, a heroin addict robs a convenience store and in a crime gone bad ends up killing the owner.

A serial killer rapes and strangles multiple victims before finally being apprehended.

At his dying wife's request, an elderly husband takes her life by giving her sleeping pills and smothering her with a pillow.

A mother leaves her infant son unattended for several hours in a steaming parked car with the windows rolled up. He dies.

Mistaking her postman for a messenger of Satan, a woman shoots him dead at her front door with a double-barreled shotgun.

After months spent planning his revenge for imagined wrongs, a 12-year-old boy guns down two of his classmates.

Although some of these cases would require highest security containment, others, after a preliminary evaluation, might well be rehabbed outside prison or jail. In making such challenging decisions, future judges and juries likely will make use of specialists and evolving artificial intelligence. In his fascinating book, *AI Super-Powers: China, Silicon Valley, and the New World Order*, Kai-Fu Lee discusses a Chinese AI company, iFlyTek, that specializes in an AI program designed to aid in courtroom decisions. Based on massive data analysis this program provides superior guidance to judges and juries regarding the strength of evidence and appropriate sentencing. It seems likely that in the near future there will be more advanced AI programs for individualized assessments of criminal risks and optimal elements of corrective education that can be used to carry out post-blame/punishment criminal justice (Kai-Fu Lee, 2018).

Some will say such an approach carries too great a risk of miscalculation. But consider what goes on now. Most violent criminals are eventually released having "done their time" only to commit violent acts again. Others who commit an extremely violent act but subsequently become low-risk persons continue to be treated as high risk, sometimes for the rest of their lives.

Official estimates characterize approximately half of persons in our country's state prisons as violent, but these figures are grossly misleading. In many states "violent felonies" include crimes such as breaking and entering an empty house or snatching a smartphone from someone on the street (Bazelon, 2019a). Still, even with a blameless approach, whatever the actual number of truly violent persons, they will create a special challenge for determining how long they will need to be contained and at what level of security. Unfortunately, perfect prediction of future dangerousness has long eluded the law and behavioral scientists. Even with the future addition of AI programs, errors will be made. One alternative would be to err on the side of caution and imprison all persons who commit violent crimes for the rest of their lives. Such an approach seems overly draconian, expensive, and immoral.

Darnell Epps was 29 years old, his brother 21 when they went after a gang member who a few days earlier had sexually assaulted the brother's wife. The young man died of gun-shot wounds, and the Epps brothers ended up in a maximum security prison in upstate New York (Epps, 2018). Under the guidance of some "older-timer" inmates who made it their responsibility to try and steer younger inmates away from trouble in the crime-breeding confines of mass incarceration, the brothers thrived. As a result they eventually created their own monitoring program for younger inmates ages 16-21. Seventeen years later (minimal time required) the brothers were paroled, unusual at the time. (Tougher sentencing laws have resulted in more than 160,000 prisoners over the age of 55 being held prisoners in state and federal prisons, regardless of what kind of person they become.)

Epps himself has argued for a different approach. "It's clear," he insists, "that our prison population is aging, but we cannot die ourselves out of mass incarceration... We must seriously consider whether society would benefit by letting reformed offenders reenter their community and whether it's economical and humane to punish solely for the sake of punishment." (Note: Epps makes his case without any reference to the immorality of blame itself.)

In February 2018, 42-year-old Tami Joy Huntsman pleaded guilty to fatally torturing (first-degree murder) her niece and nephew—Delylah Tara, age 3, and Shaun Tara, age 6. The children had been starved and beaten while staying with Huntsman and her boyfriend after the mother was killed in a car accident. With an arrangement plea agreement, Huntsman avoided the death penalty, receiving instead a sentence of life in prison without possibility of parole or appeal (Adami, 2018; Cortez, 2018). In our current system, this woman's admittedly horrific crime led to a full-on, blame-based judicial outcome. With no exceptions, she will remain in prison the rest of her life, regardless of any rehabilitative progress she might make. In contrast a blameless approach, while fully acknowledging the tragic nature of this crime, would keep this woman out of society only as long as she was a significant risk for committing such violence again. During her confinement whatever skills she possessed would be put to good use and further developed. She would be provided appropriate treatment, rehabilitative services, training, and education in a humane setting. Depending on what kind of person she became, her confinement would vary in length, and if or when she was released she would be monitored even longer. The point is this: she wouldn't necessarily stay in prison for the rest of her life solely as a matter of punishment and trying to guarantee 100% she would never kill again.

Violent Forever

There are levels of violence so severe that security (as opposed to punishment) even in a blameless environment will become a highly specialized endeavor. Persons involved in gang murder, serial killing, sadistic rape, and other violent crimes will require highest security for an extended period if not for the rest of their lives. Such persons must be contained to whatever extent necessary to protect others and themselves. Even so, these individuals—victims of circumstance as well as violent criminals—will not trigger inhumane measures. Even if it's for the rest of their lives, a blameless version of criminal justice will offer these individuals as satisfying and productive a life as possible, never forgetting their demonstrated capacity for violence to staff and other rule breakers.

What about the most sadistic, violent persons? Take Ted Bundy for example. Bundy as blameless? Granted, it's a stretch.

Under the guise of being a smart, good-looking, articulate and charming man, he was a deranged, cold-blooded killer. Ann Rule, a true-crime pioneer writer, unknowingly became his friend while she was documenting the hunt for the serial killer he turned out to be. Later she wrote a chilling account—*The Stranger Beside Me*—of this real-life encounter. Bundy once said of himself: "I'm the most cold-hearted son of a bitch you'll ever meet." One member of his last defense team agreed. "Ted was the very definition of heartless evil," she said.

His life showed it. A law student at one point, Bundy became a sadistic serial killer, rapist, kidnapper, necrophile, and burglar. After a decade of staunch denials, shortly before he was put to death, he confessed to 30 homicides in seven different states over four years. He admitted decapitating 12 victims and claimed he kept several of the heads in his apartment as mementos. After staging two dramatic prison escapes, he committed three more murders while on the run. Eventually, Bundy received three death sentences in two separate trials; but even as this was going on, he managed to talk a young woman into marrying him, and they had a child together. Finally, in 1989 he was executed in an electric chair at Florida State Prison. Extreme yes, but unfortunately there are those persons who for reasons beyond comprehension *are* extremely dangerous. A blameless approach to criminal justice can never ignore this fact. *Highest security is required not because vicious people are more blameworthy, but because they are more dangerous.* In such instances, the rule will be: the degree of containment necessary to eliminate the risk of serious violence, rendered as humanely as possible. By acknowledging extreme violence as a reality separate from the world of the vast majority of rule breakers, the blameless/responsibility approach encourages ways of dealing with it separate and distinct from what is appropriate for most rule breakers. Our current system of criminal justice is clumsy in making this distinction.

Resistance to Change

As it is, in U.S. prisons security remains the number one consideration. Mass methods predominate with relatively little attention given to individual needs. Other services such as training, treatment, education as well as preparation for reentry are secondary concerns. And why not? If punishment and escape prevention are the highest priorities, why allow resources to be siphoned off for anything else, particularly when it can be interpreted as going soft on crime and rewarding persons who deserve nothing but punishment.

Rampant prison recidivism is standard fare, stemming mainly from an absence of serious efforts to help rule breakers become social and job ready. With a major redesign, what in the past have been punishment-and-discipline centers could become cutting edge rehabilitative education/training programs aimed at providing persons with marketable skills. Mutually beneficial arrangements could be worked out with companies who badly need skilled labor and would be willing to provide resources for job training and guarantees of job placement after prison.

As a sizeable percentage of our nation's population enters a glide path to retirement, we face the challenge of having enough workers with special skills. Although it's unlikely all persons seen in the criminal justice system would be appropriate for such programs, we won't know without proper assessments and a much greater emphasis on job-ready training programs.

As long as "punishment" remains the core principle of prisons and jails, the *status quo* will persist. Undoubtedly, powerful special interests will lean against major modifications; after all, prisons have been this way for quite some time, and the practices now in place are doing a good job of *keeping them full*. This built-in resistance combines with the absence of any great public support for tackling problems of marginal individuals who don't fit easily into society, especially if they are serious rule breakers. Far easier to keep them out of sight, doing as little as possible to break their vicious cycles. And why should taxpayers pay for rehabilitative job training for rule breakers in prison? It's a legitimate question.

The most compelling answer has to do with prevention. Our current approach is rife with return trips, all of which are extremely costly. If a program truly reduces recidivism and substitutes less expensive alternatives for costly incarceration, likely it will generate substantial savings for use in new programming; and, it would not be out of the question to require reasonable payback from reformed rule breakers who become gainfully employed as a result of special training and education in prison.

End of the Blame Game

Without benefit of modern science, over a century ago a famous jurist predicted the ultimate rejection of blame. He noted how in ancient law common *objects* were sometimes mistakenly tried in court as willful agents and found deserving of punishment. So if the wheel of a runaway cart crushed a man, the *evil* wheel was blamed for his death. (Wheels were sometimes set afire as punishment for cart accidents.) Justice Holmes saw a similar primitive logic at work in the modern assignment of blame. As criminal law matured, he reasoned, blame would come to be viewed as inappropriate for persons doing what they were compelled to do; what anyone else in their exact same shoes would have done.

It's the biological, genetic, and experiential elements interacting that are the true causative culprits of crime, but because they remain beyond our comprehension, we settle for scapegoating individual rule breakers in a kind of pseudo-moral balancing act which some would maintain—even if it's off the mark—is the only way to prevent social chaos. Wrong. Nothing about holding persons responsible *without blame* does away with reasonable consequences. Nothing about it need jeopardize public safety. The only thing jettisoned with this approach is punishment for the sake of punishment which as we have seen derives its only justification from an illusion.

Even without reference to the questionable nature of blame, the case for radically changing the way our society deals with criminals is compelling. As it stands, our criminal justice system is inefficient, inhumane, extremely costly, and a perpetrator of more violence and crime. The neurobehavioral evidence we have reviewed only adds to the case against our current approach by revealing that it is also immoral.

Belief in self-agency and self-determination has had a long run. Arising from a deep-seated, but totally misguided conviction, it has long masqueraded as the crowning trait of human life: the free-willing self. It is a belief whose time is passing. As science relentlessly continues to further define the illusionary nature of self and the blame that goes with it, vengeful punishment loses its moorings, and the blame game is revealed for what it is.

Chapter 11

Conclusion

The world makes much less sense than you think. The coherence comes mostly from the way your mind works.
Daniel Kahneman

Several centuries ago French philosopher René Descartes missed the mark when he located the mind/soul in the pineal gland where, so he claimed, it was guided by God through mysterious wave actions. But Decartes' more general conclusion that we live two separate realities was right on target: one material, the other emergent and subjective. In retrospect Descartes' revelation cracked open the door to an alternative view of the self. Now neurobehavioral scientists have walked through it to provide a more detailed accounting, one that makes clear: we are not who we think we are. Not even close.

While our physical bodies operate in a *material reality*, one we can measure, monitor, and calculate, it's only after input makes its way through sensory receptors, relay centers, and higher brain processes that it is finally translated into something entirely different: a subjective reality that gives rise to the human experience. Neuronal electrochemical signals become blue sky and moving music. As counterintuitive as this seems, this subjective reality translation is all we know directly. Material reality—all aspects including the workings of the brain and the physical world around us—comes to us indirectly via translation courtesy of a deceptive brain.

Why subjective reality emerges, we can only speculate, but one distinct possibility is simplification. The complexity of the material brain with its massive tangle of neurons processing billions of bits of information each second likely would be overwhelming to our consciousness absent a far simpler version. As a stand-in, subjective reality in the form of a metaphorical narrative provides a selective awareness and orientation to the world around us as to what has happened, is happening, and may happen in the future. What is sacrificed in the way of detail is made up for by ease of understanding and unrestrained narrative imagination.

While this brain translation can be viewed as highly creative, there is a problem. This realer-than-real subjective world is an illusion. The implications are far reaching. Take the matter of making up our minds. Decisions seemingly made by our conscious selves are in fact brain echoes rendered in an after-the-fact translation as a narrative. Although in this story the chief protagonist—each of us—is experienced as a free-willing, choice-making, super action figure, with respect to material reality, the *self* makes *no* choices and initiates *no* actions. What it does do is anchor an illusionary storyline that gives meaning to our lives and orients us to the material world around us.

On the surface this revised view of the self seems absurd. And deflating. Are our lives nothing more than fixed renderings of fate told in story form? The answer is no. It's not that choices aren't made in our lives. Our brains are constantly selecting one course of action over another. If humans are anything, we are adaptive. This trait has carried us a long way as choices are being made all the time. The illusion we experience concerns *how and when these choices are made*. Despite how we experience them, they are not *without-cause* happenings springing spontaneously from our minds but rather the product of a myriad of interacting genetic and experiential influences occurring outside our awareness.

The sense of choice "we" experience is the work of a brain exerting poetic license as it crafts a metaphorical, summarizing narrative. Such is the nature of the astonishing illusion neurobehavioral science reveals and continues to elaborate. As counterintuitive as it seems, the course of our lives is strictly determined. It's only in the translated story we experience ourselves as free of the restraints of cause and effect. We only feel we are going to get out of bed and then after the fact attribute our action to what we felt. True freedom comes from no longer being tied to this illusion.

Even so, it's this subjective reality portrayal of us as free-willing, free-choosing entities that informs the practice of blaming and punishing. In light of these neurobehavioral science revelations it's clear our criminal justice system is badly misinformed. Our current practices of blaming and punishing are more akin to prosecuting a character in a novel who commits a crime than to rendering justice.

Although we are saddled with our personal feelings of blame and guilt (it's the way we are wired), when confronted by substantial evidence to the contrary, as individuals and as a society we should be able to intellectually *override* this illusion similar to the way we dismiss illusions we experience at the hands of great magicians and nature itself as we misperceive the moon "rising" and "setting."

Some will maintain that without blame and punishment society would quickly devolve into lawless chaos. In doing so they mistakenly equate blame with responsibility. The two are not the same. Blame assumes intention and choice. Responsibility does not. In a modern world this is well illustrated by no-fault laws which operate without the implication of blame. You park your car in a restricted area. You face the consequence. You pay the fine. No questions asked. No blame or punishment required. Based on what we have reviewed in this book: as social beings, even though no one is ever to blame for anything, all persons must be held responsible with consequences absent punishment.

Without the requirement of responsibility our world would become chaotic. Punishment is a different story. In fact punishment in the absence of blame is immoral, and there is no reason why responsibility cannot be fully addressed *without* punishment. Despite what our current criminal justice system implies, the two need not be tied together. Still, there will be those who insist that until we know every detail of why people do what they do, blame and punishment must be preserved. Isn't it possible, they will assert, somewhere in the workings of the brain is a yet undiscovered, quantum-like, free-willing element that does make us blameworthy. At best this seems a feeble, temporizing-and-grasping-at-straws argument. Although we may never know the precise molecular details of what is behind every translation into human emotion, thought, or act, in a strictly determined world we can presume the genetic/experiential causes are there—most of them operating *outside our conscious awareness*. To believe otherwise is to imbue willful imagination with powers not otherwise found anywhere else.

What for centuries has been a standoff debate among philosophers and theologians, now with neurobehavioral science's entry into the discussion becomes heavily weighted *against free will, choice, and blame.* For the law and courts to ignore this growing body of evidence in deference to a long-standing errant popular belief would be a dereliction of duty; a glaring injustice.

But what about deterrence? There will be those who insist, blame or no blame, punishment must be preserved for its *deterrence* value. Given the off-the-chart high rates of recidivism racked up by our current punishment-centered correctional system, it's hard to take this assertion seriously. If anything, numerous studies show punishment-oriented imprisonment actually increases the risk of more criminality. Some criminologists have argued in rebuttal that if punishment were handed out consistently and in a timely fashion, things would be different. Then punishment would have a deterrent effect. Maybe, maybe not, but with the inherent delays in our civil rights-oriented judicial system and the persistent errors in judge and jury verdicts, it seems highly unlikely this brand of punishment would ever be possible.

Currently, our criminal justice "correctional" institutions are punishment centers: mass-incarceration hellholes, riddled with systematic racism, fraught with daily threats of injury or death, and devoid of individual dignity, privacy, and respect. At considerable unnecessary expense, these inhumane paramilitary programs operate as though all inmates not only are fully blameworthy but also are extremely violent and beyond rehabilitation, deserving targets of whatever extra punishment comes their way. The most notable accomplishment of this approach is the creation of cultures of danger and fear, training in criminality, exposure to an array of drugs, and sky-high recidivism. These incubators of future criminality persist relatively unchanged mainly due to their having become public symbols of being "tough on crime."

As we have seen, blame is the bedrock assumption on which they are based. Much of this book concerns itself with how blame and punishment are applied based solely on the illusion of human choice and self-determination. Absent blame, in addition to being ineffective and wasteful, these institutions and their larger context—our criminal justice system—become blatantly unjust.

An alternative criminal justice based on blameless responsibility would be an entirely different enterprise; less institutional, less paramilitary and less inhumane. Punishment for the sake of punishment would be out. Practices modeled along the lines of "strict liability" would be in. Mass incarceration would be deemphasized and strategies for protecting against violent individuals would be smarter, more efficient, and far more individualized—all of this while maintaining public safety as a highest priority. Emphasis would be on rehabilitation, appropriate treatment, meaningful education, training, and resocialization. Much of what is accomplished currently through mass incarceration would be better addressed with far less expense outside prison walls with appropriate monitoring using the best high tech has to offer. Rates of recidivism would be closely followed as indicators of success or failure. Staff advancement, program funding, and overall policy would all be tied to keeping persons from returning to the criminal justice system.

For persons deemed too risky for release, special effort would be made to help them work toward an eventual sign off and reentry. For those who require containment for the rest of their lives—sadly, there will be some—the goal would be to help them find as much meaning as possible in a humane, productive, and safe setting while at the same time protecting the public.

As for our sense of evil, we have been looking in the wrong place. The traditional view of evil targets rule breakers themselves for vengeance and retribution. But neurobehavioral science is forcing us to relocate evil *outside the self* in the acts themselves and their true causes. By doing so evil becomes more like natural catastrophes than willful products of human monsters. While this view of evil may be far less emotionally satisfying than one driven by images of demonic persons fully deserving of vengeance, it is consistent with what we are coming to know about ourselves. We can handle it.

To continue our romance with the pretender *self* as master-in-chief is to imbue willful imagination with powers it does not have. For those who resist the case against free will because it's incomplete, I would say only this: while supportive evidence for blameless responsibility continues to grow, proof of an unfettered self that operates like an undisciplined genii choosing and willing itself through life in defiance of basic laws of cause and effect—the current prevailing universal belief—remains nonexistent.

Franz Kafka, on being told by a graphologist he had

"literary interests," replied: "No. I have no literary interests. I am literature." The evidence supports him. We are all story characters emerging from the brain, embellished non-action figures in an after-the-fact, summarizing narrative revealed to us only after unconscious brain decisions have already been made. Although the intricate details of how this astonishing illusion comes about remain unknown, we already have evidence enough to put aside the deeply entrenched but absurd belief that we humans routinely defy the physical laws of the universe and miraculously free will ourselves through life making intentional choices that leave us fully blameworthy and deserving of punishment. Based on relentlessly accumulating evidence, in time a radically revised version of criminal justice will be forced. Better it happen sooner than later.

Acknowledgements

In its early form this book was a mess. That it ever became anything otherwise was the result of a few hardy souls who agreed to slog through it and offer their best ideas as to how it might be rescued from its wayward ways. For this I am deeply indebted to: Fuller and Barbara Torrey, Miriam Cotler, Jer Lynn, Betsy Arumi, Vanessa Taylor, and Dana Weston. To the degree the book is compelling and coherent, these friends and colleagues deserve a big slice of the credit.

Author Biography

Robert Taylor is a psychiatrist and author. He first became interested in the relationship between mind and body in medical school. After his training in psychiatry at Stanford University School of Medicine, it became the basis for his first book, *Mind or Body* (McGraw Hill, 1982) followed later by *Psychological Masquerade: Distinguishing Psychological from Organic Disorders* (Springer, 2000), now in its third edition. He has also published three other books—*Health Fact, Health Fiction* (Taylor Publishing, 1990), *Finding the Right Psychiatrist: A Guide for Discerning Consumers* (Rutgers University Press, 2014), and *Madhouse Blues* (Argus Books, 2017)—the last one, a novel.

He has also written over 50 articles such as: "The Culture of Bureaucracy" (*The Washington Monthly*), "The Pseudo-Regulation of American Psychiatry" (*American Journal of Psychiatry*), "Everything's Bad for You" (*Republic Magazine*), "American Presidential Assassination" (*Violence and the Struggle for Existence*), "Beware of Health Hype" (*Reader's Digest*), and "Psychological Hedging" (*Medical Opinion and Review*).

His career in psychiatry includes training specialist at the NIMH and multiple health and mental health directorships. He was formerly an Associate Clinical Professor of Family Medicine (Stanford University School of Medicine) and a consultant to the U.S. Secret Service on presidential assassination. He has testified at the request of the U.S. Congress on national health insurance and has had substantial experience working in prisons including serving as Chief Psychiatrist at Mule Creek State Prison in California and providing consultation to two other state prisons, one of them the "supermax" facility at Pelican Bay (California).

The author has unique experience in the area of health media. In the 80s working as a consultant, he helped design the

Mental Health and Health Promotion unit of the California State Department of Mental Health with its main focus on harnessing print and electronic media in the interest of wellness. He went on to serve as Executive Producer, writer, and radio voice for the PBS radio series, *STAYING WELL* (carried on 170 stations nationwide) and Executive Project Director of *FRIENDS CAN BE GOOD MEDICINE*, a California public health wellness campaign. In this capacity he was in charge of background research and collaboration with film directors, television producers, and scriptwriters as well overseeing the design of press kits for national distribution. He made two national television appearances on NBC's *Today Show* and had numerous interviews on local and national talk radio.

For the last several years, the author has spent most of his time writing and practicing psychiatry, including clinical assignments in New Zealand and Alaska.

References

Abbott, Geoffrey. *The Book of Execution: An Encyclopedia of Methods of Judicial Execution*. London: Headline Book Publishing, 1994.

AbuDagga, Azza, et al. "Individuals with Serious Mental Illnesses in County Jails: A Survey of Jail Staffs Perspectives." Washington D.C./Arlington, VA: Public Citizen/Advocacy Treatment Center, July 14, 2016.

Adami, Chelcey. "Huntsman Sentenced to Life in Prison for Murder, Torture of Children." *The Californian*, May 18, 2018.

Alexander, Michelle. *The New Jim Crow: Mass Incarceration in the Age of Colorblindness*. NY: The New Press, 2010.

Allday, Erin. "Many Scientists Wary of Quest for 'Gay Gene'." *San Francisco Chronicle*, September 2, 2019, A1, A8.

Amarillo Globe News. "Artist's Son Convicted in Mother's Death." December 11, 1999.

Amnesty International. "The Death Penalty in 2017: Facts and Figures." April 12, 2018. Available online at: <https://www.amnesty.org/en/latest/news/2018/04/death-penalty-facts-and-figures-2017/>.

Anil, Seth. "Your brain hallucinates your conscious reality." TED Talk, April 2017.

Armstrong, Ken and S. Mills. "Part 1: Death Row Injustice Derailed." *Chicago Tribune*, November 14, 1999.

Baatz, Simon. *For the Thrill of It: Leopold, Loeb, and the Murder that Shocked Jazz Age Chicago*. New York: HarperCollins Publishers, 2008.

Baharloo, S., P. Johnston, S. Service, et al. "Absolute Pitch: An Approach for Identification of Genetic and Nongenetic Components." *American Journal of Human Genetics* 62: 224-231, 1998.

Bailey, F. Lee, H. Aronson. *The Defense Never Rests*. NY: Signet, 1972.

Barshad, Amos. "Catching a Ride On the Juul Wave." *New York Times*, Sunday Styles, April 8, 2018.

Bauer, Shane. *American Prison: A Reporter's Undercover Journey into the Business of Punishment*. NY: Penguin Books, 2018.

Bazelon, Emily. "If Prisons Aren't the Answer, What Is?" *New York Times*, Sunday Review, April 2019, p. 3.

Bazelon, Emily. *Charged: The New Movement to Transform American Prosecution and End Mass Incarceration*. New York: Random House, 2019.

Bazelon, Emily, I. Rahman. "The Law Could End Mass Incarceration." *New York Times*, Sunday Review, January 26, 2020, p. 9.

Becker, Ernest. *The Denial of Death*. NY: Free Press, 1974.

Bedau, Hugo (Editor). *The Death Penalty in America: Current Controversies*. NY: Oxford University Press, 1997.

Bedau, Hugo, M. Radelet. "Miscarriages of Justice in Potentially Capital Cases." *Stanford Law Review* 40: 21, 1987.

Benforado, Adam. *Unfair: The New Science of Criminal Injustice*. NY: Crown Publishers, 2015.

Berlin, Isaiah, H. Hardy. *The Proper Study of Mankind: An Anthology of Essays*. NY: Farrar, Straus and Giroux, 1998.

Berryhill, Michael. *The Trials of Eroy Brown: The Murder Case that Shook the Texas Prison System*. Austin: University of Texas Press, 2011.

Blackmore, Susan. *The Meme Machine*. NY: Oxford University Press, 1999.

Blakeslee, Nate. "Crime Pays." *New York Times* Book Review, October 7, 2018, pp. 1, 20.

Bohn, Dieter. "What it's like to watch an IBM AI successfully debate humans." *The Verge*, June 18, 2018. Available online at: <https://www.theverge.com/2018/6/18/17477686/ibm-project-debater-ai>.

Bone, Eugenia. *Microbia*. NY: Rodale, 2018.

Bonner, Raymond. "Argument Escalates on Executing Retarded."

New York Times, July 23, 2001.

Boroff, David. "Couple Falsely Accused of Abuse, Satanic Rituals at Day Care Center Will Receive \$3.4 M After Spending 21 Years in Prison." *New York Daily News*, August 24, 2017.

Bouchard, Thomas, D. Lykken, M. McGue, et al. "Sources of Human Psychological Differences: The Minnesota Study of Twins Raised Apart." *Science*, New Series 250 (4978): 223-228, October 12, 1990.

Bradford, William. "An Enquiry How Far the Punishment of Death Is Necessary in Pennsylvania." *Journal of Legal History* 12: 122-175, 1968 (1793).

Brady, Joseph. "Ulcers in 'Executive' Monkeys." *Scientific American* 199: 95-100, 1958.

Breyer, Stephen. *Against the Death Penalty*. Washington D.C.: Brookings Institution Press, 2016.

Brooks, Megan. "FDA OKs Novel Enzyme Therapy for Rare Disease Phenylketonuria." *Medscape*, May 25, 2018. Available online at: <https://www.medscape.com/viewarticle/897246>.

Brunner, H.G., M. Nelen, X. Breakefield, H. Ropers, B. van Oost. "Abnormal behavior associated with a point mutation in the structural gene for monoamine oxidase A." *Science* 262: 578-80, 1993.

Bryson, Bill. *The Body: A Guide for Occupants*. New York: Doubleday, 2019.

Bugliosi, Vincent. *Outrage: The Five Reasons O. J. Simpson Got Away with Murder*. NY: Island Books, 1996.

Buss, David. *Evolutionary Psychology: The New Science of the Mind* (5th Edition). NY: Routledge, 2015.

Butler, Samuel. *Erewhon*. London: Penguin English Library, 1970.

Campbell, Charlie. *Scapegoat: A History of Blaming Other People*. London: Duckworth Overlook, 2011.

Campbell, Tom, H. Owhadi, J. Sauvageau, D. Watkinson. "On Testing the Simulation Theory." *International Journal of Quantum Foundations* 3: 78-79, June 17, 2017.

Chetty, Raj, N. Hendren, L. Katz. "The Effects of Exposure to Better Neighborhoods on Children: New Evidence from the Moving to Opportunity Project." *American Economic Review* 106: 855-902, 2016.

Christian, David. *Origin Story: A Big History of Everything*. New York: Little, Brown Spark, 2018.

Conis, Elena. "The Case of the Ex-Girls." *Los Angeles Times*, Special Issue: *Men's Health/Esoterica Medica*, October 16, 2006.

Cookman, Leo. *Westworld* (Television). *Philosophy Now*, Issue 120, December 2017. Available online at: <https://philosophynow. org/issues/120/Westworld>.

Cortez, Felix. "Salina Child Killer Maintains Her Innocence 1 Day After Guilty Plea." *KSBW (NOWCAST)*, March 1, 2018.

Court TV. *Mugshots: A Mother's Madness: Andrea Yates*. October 28, 2002.

Cunningham, Amy, H. Bausell, M. Brown, et al. "Recommendations for the Use of Sapropterin in Phenylketonuria." *Molecular Genetics and Metabolism* 106: 269-276, 2012.

Dallas Morning News. "Editorial: Free to Call." September 5, 2018.

Damasio, Antonio. *Descartes' Error*. NY: Avon, 1995.

Darrow, Clarence. *The Plea of Clarence Darrow: In Defense of Richard Loeb and Nathan Leopold, Jr. On Trial for Murder*. NY: Clee Books, 2017.

Darrow, Clarence. *The Story of My Life*. NY: Da Capo Press, 1996.

Davis, Don. *Bad Blood: The Shocking Story Behind the Menendez Killings*. NY: St. Martin's Press, 1994.

Dawkins, Richard. *The Selfish Gene*. NY: Oxford University Press, 2015.

Death Penalty Information Center. "Facts About the Death Penalty." June 9, 2017, Washington, D.C. (www. deathpenaltyinfo.org).

Deitch, Michele. "What's Going On in Our Prisons?" Osher Lifelong Learning Institute Lecture, University of Texas, February 13, 2018.

Dennett, Daniel. *Elbow Room: The Varieties of Free Will Worth Wanting*. Cambridge, MA: MIT Press, 1984.

Dennett, Daniel. "Why Everyone is a Novelist." *Times Literary Supplement* 4: 459, September 16-22, 1988.

Dennett, Daniel. *Consciousness Explained*. Boston: Little, Brown and Company, 1991.

Dennett, Daniel. *Freedom Evolves*. NY: Penguin Books, 2003.

Dershowitz, Alan. *Just Revenge*. NY: Warner Books, 1999.

Dershowitz, Alan. *The Abuse Excuse*. NY: Little, Brown and Company, 1994.

Dewan, Shaila. "The Violence of Prison: A Rare, and Troubling, Look Behind the Walls." *New York Times*, March 31, 2019, A18.

Dinan, Timothy, R. Stilling, C. Stanton, et al. "Collective Unconscious: How Gut Microbes Shape Human Behavior." *Journal of Psychiatric Research* 63: 1-9, 2015.

Doctor News (*Forbes* Contributor). "Just How Smart Is Smart Medicine? MIT Scientists Are About to Find Out." April 11, 2018.

Doudna, Jennifer, S. Sternberg. *A Crack in Creation: Gene Editing and the Unthinkable Power to Control Evolution*. Boston: Mariner Books, 2017.

Dwyer, Jim, Peter Neufeld, Barry Scheck. *Actual Innocence*. NY: Random House, 2000.

Eagleman, David. *The Brain: The Story of You*. NY: Pantheon Books, 2015.

Eagleman, David. *Incognito: The Secret Lives of the Brain*. NY: Vintage Books, 2011.

Eapen, V., D. Pauls, M. Robertson. "The Role of Clinical Phenotypes in Understanding the Genetics of Obsessive-Compulsive Disorder." *Journal of Psychosomatic Research* 61: 359-364, 2006.

Einstein, Albert. Speech to the Spinoza Society, 1932.

Einstein, Albert. "My Credo." Speech to the German League of Human Rights, 1932.

Epps, Darnell. "The Prison 'Old-Timers' Who Gave Me Life." *New York Times*, Sunday Review, October 7, 2018, p. 7.

Epstein, Randi. *Aroused: The History of Hormones and How They Control Just About Everything.* New York: W.W. Norton & Company, 2018, Chapter 15.

Erbentraut, Joseph. "What the U.S. Can Learn From Prison Reform Efforts Throughout the World." Impact (*HuffPost*), December 6, 2017. Available online at: <https://www.huffingtonpost.co.uk/entry/prison-reform-international-examples_n_6995132?ri18n=true>.

Estrich, Susan. *Getting Away with Murder: How Politics is Destroying the Criminal Justice System.* Cambridge, MA: Harvard University Press, 1998.

Evans, E.P. *The Criminal Prosecution and Capital Punishment of Animals: The Lost History of Europe's Animals Trials.* NY: Faber & Faber, 1987.

Evatt, Cris. *The Myth of Free Will.* Sausalito, CA: Cafe Essays, 2010.

FBI. "Crime in the US 2014, Uniform Crime Reporting." FBI. GOV, 2014.

Ferholt, Julian, M. Genel, D. Rotnem, et al. "A Psychodynamic Study of Psychosomatic Dwarfism: A Syndrome of Depression, Personality Disorder, and Impaired Growth." *Journal of the American Academy of Child Psychiatry* 24: 49-57, 1985.

Festinger, Leon. *A Theory of Cognitive Dissonance.* Stanford, CA: Stanford University Press, 1962.

Fisher, Helen. *Anatomy of Love: A Natural History of Mating, Marriage, and Why We Stray.* NY: W.W. Norton, 2016.

Ford, Martin. *Rise of the Robots: Technology and the Threat of a Jobless Future.* NY: Basic Books, 2015.

Frances, Allen. *Saving Normal.* NY: William Morrow, 2013.

Frances, Allen, M. Ruffalo. "Mental Illness, Civil Liberty, and Common Sense." *Psychiatric Times*, May 3, 2018.

Frankl, Viktor. *Man's Search for Meaning.* Boston: Beacon Press, 1959.

Frazer, James. *The Golden Bough*. NY: Oxford University Press, 1994.

Freeman, Jim, T. Turchie, M. Noel. *Unabomber: How the FBI Broke Its Rules to Capture the Terrorist Ted Kaczynski*. Palisades, NY: History Publishing Company, 2014.

FreePeopleSearch.org/blog/were-all-related-12-things-you-might-not-know-about-human-dna.html, Lizzie. "We're All Related: 12 Things You Might Not Know about Human DNA." October 4, 2013.

Friedman, Richard. "What Cookies and Meth Have in Common." *New York Times*, Sunday Review, page 1+, July 2, 2017.

Frisbie, Thomas, R. Garrett. *Victims of Justice: The True Story of Two Innocent Men Condemned to Die and a Prosecution Out of Control*. NY: Avon Books, 1998.

Gaarder, Jostein. *Sophie's World: A Novel about the History of Philosophy*. NY: Farrar, Straus and Giroux, 1994.

Gartland, F. "PKU Case Study: Ciara (9) and Luke (4) Willetts." *The Irish Times*, September 25, 2017.

Gazzaniga, Michael. *Who's In Charge? Free Will and the Science of the Brain*. NY: HarperCollins, 2011.

Gazzaniga, Michael. *The Mind's Past*. Berkeley: University of California Press, 1998.

Gefter, Amanda. "The Case Against Reality." *Atlantic Daily* (Science), April 25, 2016.

Gilder, George. *Microcosm*. NY: Touchstone, 1989.

Goffman, Erving. *The Presentation of Self in Everyday Life*. New York: Anchor Books, 1959.

Goodall, Jane. *In the Shadow of Man*. NY: Mariner Books, 2000.

Greene, Joshua, J. Cohen. "For the Law, Neuroscience Changes Nothing and Everything." *Philosophical Transactions Royal Society London B: Biological Sciences* 359: 1775-1785, November 29, 2004.

Grim, Patrick. *Mind-Body Philosophy*. Chantilly, VA: The Great Courses, 2017.

Guinn, Jeff. *Manson: The Life and Times of Charles Manson*. NY: Simon and Schuster, 2013.

Harris, Sam. *Free Will*. NY: Free Press, 2012.

Heimbuch, Jaymi. "The Incredible Science Behind Starling Murmurations (Earth Matters)." Mother Natural Network (MNN), January 9, 2014 (Internet).

Higdon, Hal. *Leopold & Loeb: The Crime of the Century*. Chicago: University of Illinois Press, 1999.

Hirstein, William. *Brain Fiction: Self-Deception and the Riddle of Confabulation*. Cambridge, MA: MIT Press, 2006.

Hoffman, Donald. *The Case Against Reality: Why Evolution Hid the Truth from Our Eyes*. New York: W.W. Norton, 2019.

Hoffman, Donald. "Conscious Realism and the Mind-Body Problem." *Mind and Matter* 6: 87-121, 2008.

Hooper, Judith, D. Teresi. *The 3-Pound Universe: The Brain—From the Chemistry of the Mind to the New Frontiers of the Soul*. NY: Dell Publishing Co., 1986.

Humbach, John. "Doubting Free Will: Three Experiments." *Pace Law Faculty Publications*, Paper 637, 2010, S.

Husock, Howard, C. Gorman. "Bring Back the Asylum." *Wall Street Journal Review*, May 19-20, 2018.

Huxley, Thomas H. *Collected Essays: Method and Results*. NY: D. Appleton, 1894/1911.

Johnson, Kevin. "Immigration Detentions Tax Prisons." *USA Today*, June 28, 2018.

Johnson, Kevin. "Nurses, Cooks Enlisted as Guards." *USA Today*, February 14, 2018.

Kahneman, Daniel. *Thinking, Fast and Slow*. NY: Farrar, Straus and Giroux, 2011.

Karst, Kenneth, L. Leonard. *Criminal Justice and the Supreme Court*. NY: Collier Macmillan Canada, 1986.

Kelly, William, R. Pitman, W. Streusand. *From Retribution to Public Safety: Disruptive Innovation of American Criminal Justice*. Lanham, MD: Rowman & Littlefield, 2017.

Kelly, William, R. Pitman. "Reinventing Criminal Justice." Osher Lifelong Institute Lecture, University of Texas, February 13, 2018.

Kennedy, Patrick, R. Maharaj. *Dahmer Detective: The Interrogation and Investigation that Shocked the World.* NY: Poison Berry Press, 2016.

Klemm, W.R. *Mental Biology: The New Science of How the Brain and Mind Relate.* Amherst, NY: Prometheus Books, 2014.

Knappman, Edward (Editor). *American Trials of the 20th Century,* "Leopold and Loeb Trial: 1924." Detroit, MI: Visible Ink Press, 1994, pp. 83-94.

Kurzweil, Ray. *The Age of Spiritual Machines: When Computers Exceed Human Intelligence.* NY: The Penguin Group, 1999.

Kurzweil, Ray. *How to Create a Mind: The Secret of Human Thought Revealed.* NY: Penguin Books, 2013.

Kwon, Diana. "Self-Taught Robots." *Scientific American,* March 2018, pp. 27-31.

Lamson, David. *We Who Are About to Die: Prison As Seen by a Condemned Man.* NY: Charles Scribner's Sons, 1935.

Lanier, Jaron. *Dawn of the New Everything.* NY: Henry Holt and Company, 2017, pp. 55-56.

Latzer, Barry. *Death Penalty Cases: Leading U.S. Supreme Court Cases on Capital Punishment,* 3rd Edition. Burlington, MA: Elsevier, 2011.

Lawrence, Eric. "SUVs Are a Major Factor in an Alarming Increase in Deaths on Nation's Roads." *USA Today,* June 29-July 1, 2018.

Lee, Kai-Fu. *AI Super-Powers: China, Silicon Valley, and the New World Order.* NY: Houghton Mifflin Harcourt, 2018.

Libet, Benjamin, C. Gleason, D. Pearl. "Time of Conscious Intention to Act in Relation to Onset of Cerebral Activity (Readiness Potential): The Unconscious Initiation of a Freely Voluntary Act." *Brain* 106: 623-642, 1983.

Liebman, James, S. Crowley. *The Wrong Carlos.* NY: Columbia University Press, 2014.

Lifton, Robert Jay, G. Mitchell. *Who Owns Death? Capital Punishment, the American Conscience, and the End of Executions.* NY: HarperCollins, 2000.

Lindell, Chuck. "Court to Hear Texas Death Row Case." *Austin American-Statesman*, A1, A6, October 30, 2017.

Lindsey, Robert. "Dan White, Killer of San Francisco Mayor, A Suicide." *New York Times*, October 22, 1985.

Liptak, Adam. "Justices Allow Execution of Inmate Who Cannot Recall His Crime." *New York Times*, A11, November 7, 2017.

Lizzie. "We're All Related: 12 Things You Might Not Know about Human DNA." *National Geographic*/Free People Search, October 4, 2013.

Levy, David. *Love + Sex with Robots: The Evolution of Human-Robot Relationships.* NY: Harper, 2007.

Lowry, Barbara. *Crossed Over: A Murder, A Memoir.* NY: Vintage Books, 1992.

Luft, Joseph, H. Ingham. "The Johari Window: A Graphic Model of Interpersonal Awareness." Proceedings of the Western Training Laboratory in Group Development, LA, California, UCLA, 1955.

Lutz, Antoine, H. Slagter, J. Dunne, R. Davidson. "Attention Regulation and Monitoring in Meditation." *Trends in Cognitive Science* 12: 163-169, April 2008.

Maeder, Thomas. *Crime and Madness: The Origins and Evolution of the Insanity Defense.* NY: Harper & Row, 1985.

Martinson, Robert. "What Works? - questions and answers about prison reform." *National Affairs* 35: 22-54, Spring, 1974.

Maslow, Abraham. *Motivation and Personality* (3rd Edition). New Delhi, India: Pearson Education, 1987.

Mayer, Emeran. *The Mind-Gut Connection.* New York: HarperCollins Books, 2016.

McAdams, Dan. "The Psychology of Life Stories." *Review of General Psychology* 5: 100-122, 2001.

McCain, John. "Vladimir Putin Is an Evil Man." *Wall Street Journal*

Review, May 12, 2012.

McClelland, Mac. "They'll Be There Till They Die." *The New York Times Magazine*, pp. 34-41, 56-57. October 1, 2017.

McHugh, Paul. *Try to Remember: Psychiatry's Clash Over Meaning, Memory, and Mind*. NY: The Dana Foundation, 2008.

McRae, Donald. *The Great Trials of Clarence Darrow: The Landmark Cases of Leopold and Loeb, John T. Scopes, and Ossian Sweet*. New York: Harper Perennial, 2010.

Meyer, Peter. *Blind Love: The True Story of the Texas Cadet Murder*. NY: St. Martin's Press, 1998.

Minsky, Marvin. *The Society of Mind*. NY: Simon and Schuster, 1987.

Minsky, Marvin. *The Emotion Machine*. NY: Simon and Schuster, 2006.

Mischel, Walter. *The Marshmallow Test*. NY: Little, Brown and Company, 2014.

Morse, Stephen. "Brain Overclaim Syndrome and Criminal Responsibility: A Diagnostic Note." *Ohio State Journal of Criminal Law* 3: 3, 2006.

Mosley, Michael. "If You Befriend Your Gut Bacteria, Could You Help Your Immune System to Thrive?" *BBC Science Focus*, February 2018.

Mukherjee, Siddhartha. *The Gene: An Intimate History*. NY: Scribner, 2016.

Naatanen, Risto, A. Gaillard, S. Mantysalo. "Early Selective-attention Effect on Evoked Potential Reinterpreted." *ACTA Psychologica* 42: 313-329, 1978.

National Geographic, ngm.nationalgeographic.com/2013/07/125-explore/shared-genes. 2013.

National Research Council. *The Growth of Incarceration in the United States*. Washington, D.C.: The National Academies Press, 2014.

New York Times Editorial. "New York Forgets Its Juvenile Lifers." *New York Times*, Sunday Review, March 25, 2018.

Norretranders, Tor. *The User Illusion*. NY: Penguin Books, 1998.

Oliphant, Baxter. "Support for Death Penalty Lowest in Four Decades." Washington, D.C.: Pew Research Center, September 28, 2016.

O'Malley, Suzanne. "A Cry in the Dark." *O, The Oprah Magazine*, February 2002.

Oppel, Richard. "Sgt. Bowe Bergdahl's Odd Journey From Victim to Criminal." *New York Times*, A13, October 24, 2017.

Pearson, Patricia. *When She Was Bad: Violent Women and the Myth of Innocence*. Toronto: Random House of Canada, 1997.

Pendleton, Devon, C. Palmeri. "'Fortnite' Sensation Turns Developer into Billionaire." *San Francisco Chronicle*, July 30, 2018, pp. D1, D3.

Penfield, Wilder. *Mystery of the Mind: A Critical Study of Consciousness and the Human Mind*. Princeton: Princeton University Press, 1975.

Petoft, Arian. "Neurolaw: A Brief Introduction." *Iranian Journal of Neurology* 14: 53-58, 2015.

Pinker, Steven. *How the Mind Works*. NY: W.W. Norton & Company, 1997.

Plomin, Robert. *Blueprint: How DNA Makes Us Who We Are*. UK: Penguin Random House, 2018A.

Plomin, Robert. "Our Fortunetelling Genes." *Wall Street Journal Review*, November 17-18, 2018B.

Pollan, Michael. "The New Science of Psychedelics." *New York Times* Review, May 5-6, 2018.

ProCon. "Felon State Voting Laws." *ProCon.org*, April 23, 2018.

Putnam, Hilary. *Reason, Truth and History*. Cambridge, UK: Cambridge University Press, 1981.

Rachlin, Benjamin. *Ghost of the Innocent Man: A True Story of Trial and Redemption*. New York: Hachette Book Group, 2017.

Radelet, Michael, T. Lacock. "Do Executions Lower Homicide Rates: The Views of Leading Criminologists." *J. Criminal Law and Criminology* 99: 489, 2008-2009.

Radelet, Michael, H. Bedau. "The Execution of the Innocent." *Law and Contemporary Problems* 61: 105-1124, 1998.

Raemisch, Rick. "Why We Ended Long-Term Solitary Confinement in Colorado." *New York Times,* October 13, 2017.

Raine, Adrian. *The Anatomy of Violence: The Biological Roots of Crime.* NY: Pantheon Books, 2013.

Rees, Martin. *On the Future: Prospects for Humanity.* Princeton: Princeton University Press, 2018.

Reilly, Steve. "Prison Violence Rises as Budgets Slashed." *USA Today,* May 4-6, 2018.

Reiss, David. *The Relationship Code: Deciphering Genetic and Social Influences on Adolescent Development.* Boston: Harvard University Press, 2003.

Reznek, Lawrie. *Evil or Ill? Justifying the Insanity Defense.* New York: Routledge, 1997.

Rheingold, H. *Virtual Reality.* NY: Simon and Schuster, 1991.

Ridley, Matt. *Genome: The Autobiography of a Species in 23 Chapters.* New York: HarperCollins, 1999.

Ridley, Matt. *The Agile Gene: How Nature Turns on Nurture.* New York: Harper Perennial, 2004.

Rimmer, Sara and R. Bonner. "Bush Candidacy Puts Focus on Executions." *New York Times,* National Politics, May 14, 2000.

Roberts, Sam. "John Thompson, Cleared After 14 Years on Death Row, Dies at 55." *New York Times,* October 5, 2017.

Robinson, Daniel. *Wild Beasts & Idle Humours.* Cambridge, MA: Harvard University Press, 1996.

Rodriquez, Tori. "Beliefs Can Trigger Asthma Attacks." *Scientific American Mind,* Behavior and Society, March 1, 2015. Available online at: <https://www.scientificamerican.com/article/beliefs-can-trigger-asthma-attacks/>.

Rose, Steven. *Lifelines: Biology Beyond Determinism.* New York: Oxford University Press, 1997.

Rosenbaum, Ron. *Explaining Hitler: The Search for the Origins of His Evil.* NY: Da Capo Press, 2014.

Rosenblatt, Josh. "Long Road Out of Jasper: A Documentary Chronicles James Byrd Jr.'s Life and Tragic Death." *Texas Observer*, July 25, 2013.

Rule, Ann. *The Stranger Beside Me*. NY: New American Library, 1989.

Russakoff, Dale, Serge Kovaleski. "An Ordinary Boy's Extraordinary Rage." *Washington Post*, July 2, 1995.

Sapolsky, Robert M. *Behave: The Biology of Humans at Our Best and Worst*. NY: Penguin Books, 2017.

Scarr-Salapatek, Sandra. "Race, Social Class, and IQ." *Science* 174: 1285-1295, 1971.

Scharf, Michael. *Balkan Justice: The Story Behind the First International War Crimes Trial Since Nuremberg*. Durham, NC: Carolina Academic Press, 1997.

Scheck, Barry, P. Neufeld, J. Dwyer. *Actual Innocence*. NY: Doubleday, 2000.

Scheiber, Noam. "Murderer, Esq." *New York Times*, Sunday Business, February 3, 2018.

Scott, Alan. "Grand Illusions and Existential Angst." *Skeptical Inquirer*, November/December 2018, pp. 51-55.

Segal, Nancy, T.J. Bouchard. *Entwined Lives: Twins and What They Tell Us about Human Behavior*. New York: Plume Books, 2000.

Seligman, Martin. *The Hope Circuit: A Psychologist's Journey from Hopelessness to Optimism*. NY: Hachette Book Group, 2018.

Sernoffsky, Evan. "S.F. Jail Stays Cut By New City Program." *San Francisco Chronicle*, May 16, 2018.

Sharot, Tali. *The Influential Mind*. New York: Henry Holt and Company, 2017.

Sheldrake, Merlin. *Entangled Life: How Fungi Make Our Worlds, Change Our Minds & Shape Our Futures*. New York: Random House, 2020.

Simon, Herbert. *Models of Man*. Cambridge, MA: MIT Press, 2004.

Simon, Jonathan. *Mass Incarceration on Trial: A Remarkable Court Decision and the Future of Prisons in America*. NY: The New

Press, 2014.

Singh, Push. "Examining the Society of Mind." *Computers and Artificial Intelligence* 22: 521-543, 2003.

Solon, Olivia. "Is our world a simulation? Why some scientists say it's more likely than not." *The Guardian*, October 11, 2016. Available online at: <https://www.theguardian.com/technology/2016/oct/11/simulated-world-elon-musk-the-matrix>.

Sontag, Susan. *Illness as Metaphor*. New York: Farrar, Straus and Giroux, 1978.

Soon, C.S., M. Brass, H.J. Heinze, J.D. Haynes. "Unconscious determinants of free decisions in the brain." *Nature Neuroscience* 11:543-545, 2008.

Squires, Nick. "Pope Benedict XVI Urged to Reopen Swiss Guards [sic] Murder Investigation." *The Telegraph*, February 14, 2018 (Internet).

Stanley, Alessandra. "Chief Guard is Killed in Vatican Along with Wife and a 2d Man." *New York Times*, May 5, 1998.

Steiker, Carol, J. Steiker. *Courting Death: The Supreme Court and Capital Punishment*. Cambridge, MA: Belknap Press (Harvard), 2016.

Stern, Lindsay. "The Divide." *Smithsonian*, July-August 2020, 30-49, 114.

Stevenson, Bryan. *Just Mercy: A Story of Justice and Redemption*. NY: Spiegel & Grau, 2015.

Stillman, Sarah. "Separated: A Fight to Keep Mothers from Being Incarcerated." *The New Yorker*, November 5, 2018, pp. 42-53.

Storr, Anthony. *Freud: A Very Short Introduction*. New York: Oxford University Press, 1989.

Suddendorf, Thomas. "Two Key Features Created the Human Mind." *Scientific American*, September 18, 2018, pp. 43-47.

Sun, Leo. "Watching Video Games a Moneymaker." *USA Today*, August 14, 2018.

Susskind, Daniel. *A World Without Work: Technology, Automation, and How We Should Respond.* NY: Metropolitan Books, 2020.

Swanson, Sady. "Colorado Man Details How He Murdered Wife, Young Daughters." *USA Today*, March 8, 2019, 6A.

Taleb, Nassim. *The Black Swan: The Impact of the Highly Improbable.* New York: Random House, 2007.

Tani, Jun. *Exploring Robotic Minds.* NY: Oxford University Press, 2017.

Tavris, Carol, E. Aronson. *Mistakes Were Made (But Not By Me).* NY: Houghton Mifflin Harcourt, 2015.

Taylor, Robert. *Psychological Masquerade: Distinguishing Psychological from Organic Disorders*, Third Edition. NY: Springer Publishing Company, 2007.

Taylor, Robert, E.F. Torrey. "The Self-Education of Psychiatry Residents." *American Journal of Psychiatry* 128: 1116-1121, 1972.

Thompson, Heather. *Blood in the Water: The Attica Prison Uprising of 1971 and Its Legacy.* NY: Vintage, 2017.

Thomson, Helen. "Lessons from Strange Brains." *Wall Street Journal Review*, June 30/July 1, 2018, C1-C2.

Tolson, Mike. "Doctor's Effect on Justice Lingers. Testified in Many Death Row Cases." *Houston Chronicle*, June 17, 2004.

Topol, Eric. "The A.I. Diet." *New York Times*, Sunday Review, March 3, 2019.

Torrey, E. Fuller, A.D. Kennard, D. Eslinger, R. Lamb, J. Pavle. "More Mentally Ill Persons Are in Jail and Prisons than Hospitals: A Survey of States." Arlington, VA: Treatment Advocacy Center, May 2010.

Treatment Advocacy Center. "Serious Mental Illness (SMI) Prevalence in Jails and Prisons." Arlington, VA: Treatment Advocacy Center, September 2016.

Tufekci, Zeynep. "We're building a dystopia just to make people click on ads." TED Talk, September 2017.

Vera Institute of Justice Center on Sentencing and Corrections. "Incarceration's Front Door: The Misuse of Jails in America (Summary Report)." February 2015.

Verhovek, Sam. "Her Final Appeals Exhausted Tucker Is Put to Death in Texas." *New York Times*, February 4, 1998.

Viereck, George. "What Life Means to Einstein." *Saturday Evening Post*, pp. 17, 110, October 26, 1929.

Vonnegut, Kurt. *Slaughterhouse-Five*. NY: Dial Press, 2009.

Walter, Grey, R. Cooper, V.J. Aldridge, et al. "Contingent Negative Variation: An Electric Sign of Sensori-Motor Association and Expectancy in the Human Brain." *Nature* 203: 380-384, 1964.

Ward, Mike, Rebecca Rodriguez. "Texas Executes Tucker." *Austin American-Statesman*, February 4, 1998.

Wegner, Daniel. *The Illusion of Conscious Will*. Cambridge, MA: Bradford Books (MIT Press), 2002.

Weinberg, Steven. *Dreams of a Final Theory: The Scientist's Search for the Ultimate Laws of Nature*. NY: Vintage Books, 1994.

Weinstein, Henry. "A Sleeping Lawyer and a Ticket to Death Row." *Los Angeles Times*, July 15, 2000.

Weinstein, Henry. "Death Penalty Foes Focus Effort on the Innocent." *Los Angeles Times*, November 16, 1998.

Wilson, Edward. *The Meaning of Human Existence*. New York: Liveright Publishing, 2014.

Wolfe, Tom. "Sorry, But Your Soul Just Died." *Independent*, February 2, 1997. Available online at: <https://www.independent.co.uk/arts-entertainment/sorry-but-your-soul-just-died-1276509.html>.

Wootton, Barbara. *Crime and the Criminal Law*. 1964. (Out of Print).

Worthen, Molly. *The History of Christianity II: From the Reformation to the Modern Megachurch* (Great Courses Guidebook). Chantilly, VA: The Teaching Company, 2017.

Wright, Lawrence. *Remembering Satan: A Tragic Case of Recovered Memory*. NY: Vintage Books, 1994.

Wright, Lawrence. *Twins: And What They Tell Us About Who We Are*. NY: John Wiley & Sons, 1997.

Wright, Robert. *The Moral Animal*. NY: Pantheon Books, 1994.

Wright, Robert. "The Evolution of Despair." *TIME* Magazine,

June 24, 2001.

Wright, Valerie. "Deterrence in Criminal Justice: Evaluating Certainty vs. Severity of Punishment." Washington, D.C.: The Sentencing Project, November 2010.

Wright, William. *Born That Way: Genes, Behavior, Personality*. New York: Routledge, 1999.

Yong, Ed. *I Contain Multitudes: The Microbes Within Us and a Grander View of Life*. NY: HarperCollins, 2016.

Young, Dudley. *Origins of the Sacred: The Ecstasies of Love and War*. NY: Harper Perennial, 1992.

Zimmer, Carl. *She Has Her Mother's Laugh: The Powers, Perversions, and Potential of Heredity*. New York: Dutton, 2018.

Zuroff, Efraim. *Operation Last Chance: One Man's Quest to Bring Nazi Criminals to Justice*. NY: Palgrave Macmillan, 2009.

ACADEMIC AND SPECIALIST

Iff Books publishes non-fiction. It aims to work with authors and titles that augment our understanding of the human condition, society and civilisation, and the world or universe in which we live.

If you have enjoyed this book, why not tell other readers by posting a review on your preferred book site.

Recent bestsellers from Iff Books are:

Why Materialism Is Baloney
How true skeptics know there is no death and fathom answers to life, the universe, and everything
Bernardo Kastrup
A hard-nosed, logical, and skeptic non-materialist metaphysics, according to which the body is in mind, not mind in the body.
Paperback: 978-1-78279-362-5 ebook: 978-1-78279-361-8

The Fall
Steve Taylor
The Fall discusses human achievement versus the issues of war, patriarchy and social inequality.
Paperback: 978-1-78535-804-3 ebook: 978-1-78535-805-0

Brief Peeks Beyond
Critical essays on metaphysics, neuroscience, free will, skepticism and culture
Bernardo Kastrup
An incisive, original, compelling alternative to current mainstream cultural views and assumptions.
Paperback: 978-1-78535-018-4 ebook: 978-1-78535-019-1

Framespotting
Changing how you look at things changes how
you see them
Laurence & Alison Matthews
A punchy, upbeat guide to framespotting. Spot deceptions and
hidden assumptions; swap growth for growing up. See and be free.
Paperback: 978-1-78279-689-3 ebook: 978-1-78279-822-4

Is There an Afterlife?
David Fontana
Is there an Afterlife? If so what is it like? How do Western ideas
of the afterlife compare with Eastern? David Fontana presents the
historical and contemporary evidence for survival of
physical death.
Paperback: 978-1-90381-690-5

Nothing Matters
a book about nothing
Ronald Green
Thinking about Nothing opens the world to everything by
illuminating new angles to old problems and stimulating new
ways of thinking.
Paperback: 978-1-84694-707-0 ebook: 978-1-78099-016-3

Panpsychism
The Philosophy of the Sensuous Cosmos
Peter Ells
Are free will and mind chimeras? This book, anti-materialistic but
respecting science, answers: No! Mind is foundational
to all existence.
Paperback: 978-1-84694-505-2 ebook: 978-1-78099-018-7

Punk Science
Inside the Mind of God
Manjir Samanta-Laughton
Many have experienced unexplainable phenomena; God, psychic
abilities, extraordinary healing and angelic encounters. Can
cutting-edge science actually explain phenomena
previously thought of as 'paranormal'?
Paperback: 978-1-90504-793-2

The Vagabond Spirit of Poetry
Edward Clarke
Spend time with the wisest poets of the modern age and of the
past, and let Edward Clarke remind you of the importance of
poetry in our industrialized world.
Paperback: 978-1-78279-370-0 ebook: 978-1-78279-369-4

Readers of ebooks can buy or view any of these bestsellers by
clicking on the live link in the title. Most titles are published in
paperback and as an ebook. Paperbacks are available in traditional
bookshops. Both print and ebook formats are available online.
Find more titles and sign up to our readers' newsletter at
http://www.johnhuntpublishing.com/non-fiction
Follow us on Facebook at
https://www.facebook.com/JHPNonFiction
and Twitter at https://twitter.com/JHPNonFiction